AF698590

WIRELESS SENSOR AND ROBOT NETWORKS

From Topology Control to Communication Aspects

From Topology Control to Communication Aspects

Editors

Nathalie Mitton | David Simplot-Ryl

Inria Lille – Nord Europe, France

World Scientific

NEW JERSEY • LONDON • SINGAPORE • BEIJING • SHANGHAI • HONG KONG • TAIPEI • CHENNAI

Published by

World Scientific Publishing Co. Pte. Ltd.
5 Toh Tuck Link, Singapore 596224
USA office: 27 Warren Street, Suite 401-402, Hackensack, NJ 07601
UK office: 57 Shelton Street, Covent Garden, London WC2H 9HE

British Library Cataloguing-in-Publication Data
A catalogue record for this book is available from the British Library.

WIRELESS SENSOR AND ROBOT NETWORKS
From Topology Control to Communication Aspects

Copyright © 2014 by World Scientific Publishing Co. Pte. Ltd.

All rights reserved. This book, or parts thereof, may not be reproduced in any form or by any means, electronic or mechanical, including photocopying, recording or any information storage and retrieval system now known or to be invented, without written permission from the publisher.

For photocopying of material in this volume, please pay a copying fee through the Copyright Clearance Center, Inc., 222 Rosewood Drive, Danvers, MA 01923, USA. In this case permission to photocopy is not required from the publisher.

ISBN 978-981-4551-33-5

Printed in Singapore by Mainland Press Pte Ltd.

Preface

Nathalie Mitton and David Simplot-Ryl

Inria Lille – Nord Europe

Wireless sensor networks have gained much attention these last years thanks to the great set of applications they allow and technological advances. One of these challenges is node mobility: sensors could be moved unexpectedly if deployed in an uncontrolled environment or hold by moving object/animals. Beyond all this, a new dimension arises when this mobility is controlled, i.e., if these sensors are embedded in robots. These robots cohabit with sensors and cooperate together to perform a given task collectively by still presenting hardware constraints: they still rely on batteries; they communicate through short radio links and have limited capacities.

The set of potential applications of wireless sensor and robot networks is very wide. We can divide the applications into three categories:

- fleet of wireless robots,
- wireless sensor network serviced by mobile wireless robots,
- wireless sensor and robot networks.

Fleet of wireless robots

In such applications, the network is composed of a fleet of wireless robots that need to cooperate to fulfill a given task. Robots constitute the nodes of the network. The tasks may greatly differ regarding the application but the most current envisaged ones today are operations of coverage, exploration and substitution networks.

Coverage

Coverage is one of the first fundamental goal of wireless sensor networks. Due to the large variety of sensors and applications, coverage is subject to a wide range of interpretations. For example, let's consider a wireless sensor network deployed in a forest for fire detection. One may ask how well the network can observe a given area and what the chances are that a fire starting in a specific location will be detected in a given time frame. Furthermore, coverage formulations can try to find weak points in a sensor field and suggest future deployment or reconfiguration schemes for improving the overall quality of service. The quality of coverage will thus depend on sensor relative location one to each other. For a given number of sensors, the best coverage will be achieved when the monitoring area of every sensor overlaps as less as possible. At the same time, sensors should be geographically close enough to ensure network connectivity.

The use of mobile robots instead of sensors allows the improvement of such a coverage. Indeed, in a plain wireless sensor network, sensors are generally deployed at random or based on a previous off-line computation that does not take the obstacles and ground constraints into account. Using mobile robots allows an adaptive and real time deployment that better fits the environment dynamics. Robots could self-deploy to cover and then monitor an area through their embedded sensors. They thus become the network nodes. An alternative is that they are sent to dynamically drop wireless sensors at specific positions. The choice between these options will be lead by the application. Do we need to send robots because the environment is highly dynamic and thus, to achieve a good coverage, they have to continuously re-position? Is the area to monitor unaccessible or dangerous to humans? In these latter cases, robots could be sent to deploy a wireless sensor network by dropping sensors on positions dynamically computed based on local features.

Coverage can concern several kinds of coverages: area coverage, barrier coverage or simply a point of interest coverage. Chapter 4 will detail the way these issues are technically addressed in the literature.

Exploration

These issues are detailed in Chapter 7. In such applications, a fleet of wireless cooperative robots is sent for exploration. Robots are usually used when sending human is not possible. This is for instance the case on other planets like the Curiosity robot on Mars planet or after/during a disaster.

Firemen send flying drones to supervise forest fire and better anticipate the fire propagation[1]. Today, all these exploration tasks are performed by a single robot that can be powerful and thus expensive. Next steps are to send a complete fleet instead in which robots are smaller, more manageable and self-cooperative. This will allow a quicker, more reliable and cheaper exploration. Also, using heterogeneous robots in such an application allows a step further than the plain exploration as the Swarmanoid[2] European project shows. It illustrates an exploration task in which a fleet of drones is in charge of locating a book. Once the book is localized, drones guide small ground robots to it. These small ground robots cooperate to carry a climbing robot which is able to retrieve the book in a high shelf. This is a simple application which gives an overview of the wide range of possibilities that a wireless mobile robot network can achieve.

Substitution networks

Chapter 5 gives a deep insight in substitution networks and the challenges they leverage. A substitution network is a temporary network that will be deployed to support a base network in trouble and help it to provide best service. When a failure in a primary network occurs, the fleet of wireless robots self-deploy to fulfill this network application with best QoS, *i.e.*, by position themselves not only to relay the message but to relay the messages with the best link quality. To do so, robots continuously measure the signal strength and move to the position that provides the best signal.

Wireless sensor network serviced by mobile wireless robots

The second category of applications of the mobile wireless sensor and robot networks in when a sensor network is serviced by robots. The idea here is that there exists an operational wireless sensor network that requires sporadic help/servicing from mobile robots. In this category, sensors and robots perform independently.

The first application to be highlighted is the node deployment. Indeed, if there is a need to monitor a specific area through the use of wireless sensors and that this area is unreachable by humans, a fleet of mobile robots can be sent instead to drop sensors on the area. Please refer to Chapter 4 for more details.

[1] http://www.bfmtv.com/planete/un-micro-drone-aide-les-pompiers-a-lutter-contre-les-incendies-220860.html

[2] http://www.swarmanoid.org

Then, once the network is deployed, mobile robots can bring new services. Some of them could be continuous like for instance providing recharge to nodes. The idea is that mobile robots travel around sensors to reload their energy, such allowing no network death. Such an issue is discussed in Chapter 6. The challenge here is to provide enough energy to every sensor node to ensure the whole network proper functioning. Several parameters have to been considered like the recharge time, the rate of energy to provide to sensors at once (the more energy, the longer the recharging time and the less sensors recharged).

Another examples of continuous services that could be provided by robots to sensors is localization and data gathering. In the former, the idea is that robots are aware of their real time position and travel close to every sensor in the network to share this information with it (See Chapter 9). Depending on the network dynamics, such a service could be done only once or coupled with the recharging model. In the latter, autonomous robots can travel over the area to be monitored to gather information from sensor nodes (See Chapter 2). These sensor nodes could be either fixed or mobile. In such a case, the purpose is to allow sensors to empty their memory and collect more monitoring information without lost. Robots have thus to be efficient enough to self-organize and path plan such that every sensor is visited regularly and on time.

One of the most interesting example of kind of services that could be brought by wireless robots to wireless sensors is on-demand ones. Such services mostly require real-time, which leverage a new challenge. In addition to gathering and reporting data from the environment, sensors may also report failures of neighboring sensors or lack of coverage in certain neighborhood. Such event information may be gathered by nearby mobile robots through multi-hop paths. Once an event has been detected, robots coordinate with each other to make a decision on the most appropriate way to perform the action. The main such applications are intruder tracking and fire detection and limitation. In the latter case, sensors trigger an alarm upon fire detection that is routed to robots. Robots cooperate and self-organize to reach the fire point and try to confine the fire by moving around it or if they can to deaden it while waiting for firemen. Issues related to such applications are addressed in Chapters 3 and 8.

In the intruder detection application, sensors detect an unusual intrusion and report it to robots and central entity. Robots speed up to the intrusion point, locate him/her based on data collected by the wireless sensors and track him/her. Note that such safety application is well on demand currently[3].

Wireless sensor and robot networks

This section gives a quick overview of what a wireless sensor and robot network can bring if all devices cooperate at the same level, taking heterogeneity features into account at each decision step and improving locally all performances. The basic applications could be the ones as described in Chapters 4 and 1 in which the robots are part of the wireless sensor network and act only to improve the performances of this latter like routing or coverage.

Wireless sensor and robot networks can be used for rescue operations. Imagine a disaster such as the Fukushima nuclear power plant that was damaged in 2011 by an earthquake and a tsunami. Human workers are better not to be sent to measure the damages because of radiations. Imagine that a fleet of robots is sent instead. Robots will cooperate to explore the area efficiently without interfering. They have to move in a cooperative and coordinated way to maintain communication links between them. Robots will evolve in an autonomous dynamic way in this hostile area taking decisions based on environmental information. These local data, either physical if directly sensed from their environment (such as a radiation rate) or from the other robots (for cooperation and communication purposes) will drive their actions and movements. They may drop static sensors along their paths for ensuring a communication link between several area points, or to ensure a continuous monitoring while they move ahead. These sensors will be later re-used for further explorations and accurate monitoring comparisons. How has the radiation level evolved since last exploration in this accurate place?

Similarly, wireless sensor and robot networks can be used for regular place monitoring and switch to an emergency situation by simply adapting to the events they sense and changes in environments. Let's illustrate such an application for museum safety and security. For this purpose, sensors are deployed in the museum to measure specific values such as temperature, humidity, carbon concentration, that impact the art major works preservation. Robots patrol to gather data from remote sensors or to change sensor batteries. In normal use, data collected by sensors and robots are normally sent to database for daily monitoring. If an anomaly is detected

[3]http://www.ocdelibrairie.org

by the sensors, it has to adapt. If an intrusion is detected after the museum closure, sensors triggers an alarm, robots will locate and possibly surround the intruder guided by sensor data and protect art major works located and identified thanks to sensors. If a fire or a flood is detected, sensors will guide robots that will shut up endangered areas, guide evacuated people to the exit via a safe way and guide rescue operations to victims. Sensors will also allow robots to identify high value art major works, isolate, evacuate or protect them. Then, robots will explore the damaged area dropping sensors to ensure a continuous monitoring.

Challenges

Many challenges arise due to the introduction of the controlled mobility. All these challenges many depend on the application requirements and expectations but globally, some of them are common to all of them. First, where and when a robot should move? How does it cooperate with the other robots of its fleet and coordinate to fulfill a given task without being in each other's way and with minimum energy? And one of the most important: the energy? How to preserve it as much as possible while ensuring a high quality service? When and who to reload and how much? Some of these challenges are discussed in the different chapters of this book and in particular in Chapter 10.

Energy is the most challenging issue. Indeed, before performing an action, a robot has to ensure that it will have enough energy to come back to its base station for reloading, except if the application tolerates to loose robots, which is rare because of the robots cost. The energy consumption is mostly due to robot communication and movement. Algorithms should thus consider in their design these consumption sources in order to minimize them by fitting with shorter path length and less communications. Then, it is strongly connected with hardware and battery discharge and reloading. How long does a robot need to recharge fully, half, etc. should be considered as well.

Then, task assignment is the second most challenging issue. It consists in selecting robots that will be sent together by assigning them specific distributed tasks such that the fleet at the whole will act in a coordinated and efficient way. Robots have to be selected carefully still with regards to energy consumption and environment. Should the algorithm select the closest robots? Or the ones with the higher level of energy? How many?

Some aspects of task assignment may require real-time or synchronization, short wireless communications, etc. Some solutions considering all these aspects jointly and reduce action redundancy should be investigated.

Globally, wireless robot and sensor networks need a unified framework in order to constitute a flat network with different services and heterogeneous components. Several cross-layer integration issues among the communication layers should be integrated in order to improve the global efficiency and performance of a wireless sensor and orbit network. There is still a need for real-time communication protocols for robot and sensor coordinations.

With the advance in hardware technology, all these applications are about to be deployed. There exist more and more wireless sensors and robots to be networked. We can also witness the apparition of new platforms for experimental testing like FIT[4] or Project Lab[5] that allow the improvement of software protocols allowing the autonomy of such networks.

There exist several actions around the world dealing with such issues. This is for instance the case of the following projects COGX[6], ROBOTCUB[7] or PACO-PLUS in which cognitive robotics approaches are used to make the robot self-understand and self-extend its abilities. SPARK2[8], IWARD[9], ROBOSWARM[10] and SENSOPAC[11] analyze cognitive collaborative models inspired by swarm behavior or for emerging cooperative robot behavior. COMPLACS aims at developing a unified toolkit for intelligent control in many different problem areas. This toolkit will incorporate many of the most successful approaces to a variety of important control problems within a single framework, including bandit problems, Markov Decision Processes (MDPs), Partially Observable MDPs (POMDPs), continuous stochastic control, and multi-agent systems.

At the ERA (European Research Area), several collaborative projects have been performed in robotics, 15 of them involve collaboration robot-robot. Among them, some deal with micro-robots (REPLICATOR and MICRON), underwater robots (MORPH, Co3 AUVs, CoCoRo and NOPTILUS) and thus focus on a specific environment, some are devoted to flying robots (sFLY and COMETS). Others investigate various re-

[4]fit-equipex.fr
[5]http://nootropicdesign.com/projectlab/2010/03/04/wireless-robotics-platform/
[6]http://cogx.eu/
[7]http://www.robotcub.org/
[8]http://www.spark2.diees.unict.it/
[9]http://www.iward.eu/cms/index.php
[10]http://roboswarm.eu/
[11]http://www.sensopac.org/

search/application problems for ground mobile robots. For instance ARCAS, MARTHA, IWARD, DustBot and ROBOSWARM projects deal with a fixed asset of sensors for transportation, area cleaning, construction or patrolling and URUS and GUARDIANS address the aspects of sensor placing and adaptation.

Brief outline of this book

This book is composed of 10 chapters discussing about the different issues brought by a wireless sensor and robot network. It does not pretend to be exhaustive but at least to draw the perspectives and the possibilities of such networks together with the challenges they raise and how previous researches have addressed them. The first chapters show how the addition of mobility in a wireless sensor network can improve the routing performances (See Chapter 1) or the data collection (See Chapter 2). Following chapters detail the issues appearing when several robots have to cooperate to service sensors (See Chapter 3), to ensure a coverage of specific points or areas (See Chapter 4) or to support a failing existing network (Chapter 5). Chapters 7 and 8 describe alternative solutions, showing how the swarm intelligence can be used for robot cooperation and coverage area. Chapter 9 introduces the different localization techniques that could be used in such networks. Finally, Chapter 10 concludes this book by discussing challenges and business applications.

Acknowledgements

The editors are grateful to all authors for their valuable contribution to this book. We hope the readers will find the information available in this book helpful and worth reading. Every feedback from readers will be welcome.

Contents

Chapter 1

Enhancing Routing Performance in Mobile Wireless Sensor Networks through the Use of Controlled Mobility

Nicolas Gouvy[1,2], Nathalie Mitton[2] and David Simplot-Ryl[2]

[1] *Université Lille 1*, [2] *Inria Lille – Nord Europe, France*

Abstract. *Node mobility has long been considered as a hazard in wireless sensor networks, causing a degradation of performance or even persistent routing failures. The advance of technology has made possible a new vision in which mobility allows the improvement of performances and new applications. In this chapter, we focus on some of the contributions that take advantage of controlled mobility in wireless sensor networks to improve message delivery performances.*

1.1 Introduction

Wireless Sensor Networks (WSN) are sets of a handful to thousands of sensors communicating through the radio medium in a multi-hop fashion. Each sensor embeds a low-power processor with limited computing and memory capabilities, a radio device and sometimes a localization device. It is essential for the nodes to collaborate in order to route data in a reliable and energy-efficient way to a given destination. The limited energy and computing or memory capabilities of a node and the dynamic nature of wireless links impose extra difficulties in the design of efficient routing protocols. Moreover, in WSN, routing tables is subject to multiple and rapid modifications due to the dynamic nature of the topology (unreliable and unstable wireless links, mobile nodes, etc). Hence traditional network routing protocols are not compliant with energy and dynamic constraints as they rely on large and quite static routing tables stored in memory. Routing protocols for WSN are conceived with those constraints in mind tempering energy overhead, delay and complexity. A new approach in routing in WSN is to introduce controlled mobility enabled sensors. In other words, one has the possibility to relocate one or several nodes in order to modify the topology to save energy. Results in [12] show that deploying resourceful mobile devices in a WSN provides better results in terms of energy consumption than the traditional density increasing.

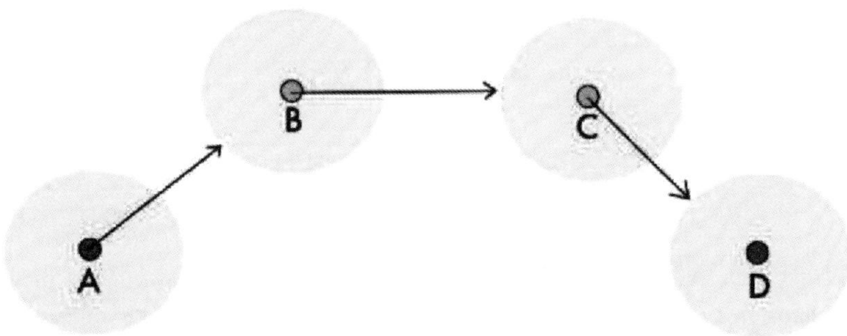

Fig. 1.1 Node A has a message for node D. A uses intermediates nodes B and C to route the message. Arrows represents node trajectories while the disk represents nodes radio range.

To the best of our knowledge, the first paper to use controlled mobility in order to improve routing performance is [8]. In this paper, authors consider the routing in a mobile network, where every node follows a different trajectory. In this kind of network, mobile nodes motions and locations are predictable or estimated with high accuracy by every node at any time for every node according to some known functions. Hence, authors propose an algorithm called **Optimal Routing Path** which makes the emitting mobile node compute a *moving routing path* over moving nodes up to the destination (Source routing). **Optimal Routing Path** is different from previous routing algorithms as it takes account of the possibility for each mobile node to deviate slightly - on purpose - from its trajectory. This temporary deviation aims to make possible message forwarding which would be impossible otherwise because of too short radio-range. Using the global knowledge of nodes movement and the possibility to modify every nod trajectory, **Optimal Routing Path** returns the shortest time strategy to route a message between two mobile nodes.

Figure 1.1 illustrates a use of **Optimal Routing Path** algorithm. The network is composed of four mobiles nodes A, B, C and D. A has a message for D which is outside its radio range. Moving directly to D would require a high energy consumption for A. Consequently, A uses its knowledge of motions at dispatch times for all network nodes to compute*the moving routing path* from A to D. According to the algorithm, A approaches B and forwards the message to it. B behaves the same towards C which moves directly to D and delivers the message.

Still, assuming that every node has a full knowledge of the location and movement of all other nodes in the network makes this approach hardly possible in wireless sensor networks in which devices are constraint in terms of memory and computing capacities.In order to overcome this strong assumption, a new routing paradigm has been proposed called Message Ferrying. Message Ferrying assumes that only some nodes are mobile and only requires the knowledge of mobile nodes.

1.2 Message Ferrying

Message Ferrying is a wireless routing paradigm in which a mobile node called *a ferry* travels inside the network following a predefined trajectory. The ferry can thus receive messages and physically carry them to another part of the network. Hence, this paradigm supposes a network where there are at least two different kinds of nodes : the message ferries (or ferry for short) and the other nodes. Although ferries and other nodes can have different hardware specifications, the differentiation criteria is that the ferry only has mobile routing purpose while the other nodes can have other activities. This paradigm differs from Opportunistic routing since Message Ferrying is proactive and non random : the message ferry will move on a defined path whatever the network activity is.

Message Ferrying has been introduced in [13] in order to improve or create connectivity in partitioned network. The ferry route design is of critical importance. It has to reduce the network latency the ferry introduces. The latency requires nodes to have enough memory to buffer the data between ferry passages. Otherwise nodes or even the ferry would have to discard data. This design is known to be an NP-Hard problem and can only be approximated.

In case of a static network such as in Fig. 1.2(a), the ferry is used to create a fully connected network. The use of a mobile ferry allows nodes

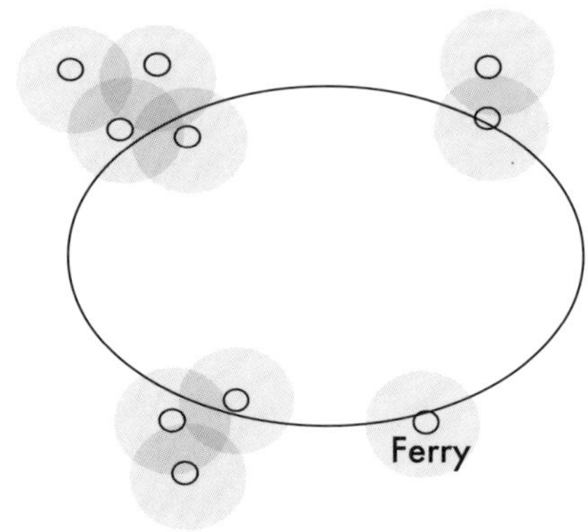

(a) Mobile Node Network

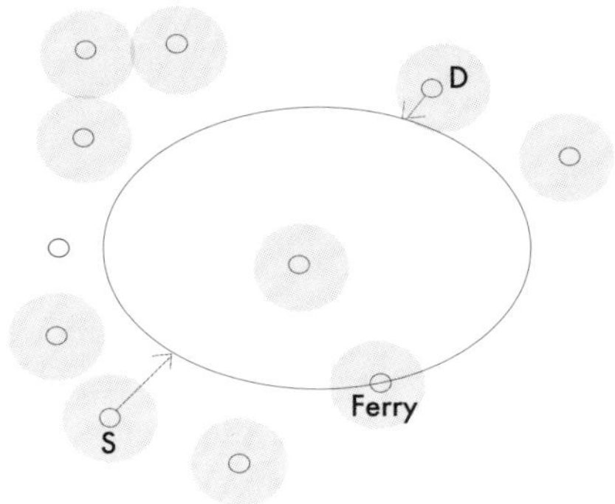

(b) Static Node Network

Fig. 1.2 Message Ferrying in action. The ferry moves accordingly to its trajectory, carrying messages physically.

to communicate with each others despite the network partition. Authors in [13] proposed an algorithm for static networks. The computation of the ferry route is made by using the traveling salesman problem (TSP) resolution algorithm over all the nodes tuned in order to reduce the total delay of the ferry route. The route is then adapted in order to satisfy bandwidth criteria. This approach makes the ferry a message carrier traveling from rendezvous point to rendezvous point. On these points, it stores messages to route, carries them to the corresponding rendezvous point before forwarding.

In mobile network, the ferry can avoid long travels for mobile nodes. This ferry-in-a-mobile-network case is addressed in [14]. In this work, the authors distinguish two different cases. The Node Initiated Message Ferry Scheme (NIMFS), which is illustrated in Fig. 1.2(b), and the Ferry Initiated Message Ferry Scheme (FIMFS).

In NIMFS, if a node S has a message for D, instead of traveling up to D, node S heads to the ferry on its trajectory and forwards it the message in a short range manner. The ferry then carries the message on its route without making changes on its trajectories. Finally, when D is detected by the ferry, the message is forwarded.

In FIMFS, when a node needs to send a message, it warns the ferry which makes a detour to join this node. In other words, the ferry actively leaves its trajectory to get the packet in a short range manner. Once the ferry has retrieved the packet, its route is recomputed in order to deliver the message. The route computation in FIMFS is one of the two following heuristics: either it always moves towards the Nearest Neighbor, or it uses a similar approach to the one in [13] computing a TSP path tuned to minimize the expected message dropping.

Numerous approaches have been proposed in order to optimize the route of the ferry in the Message Ferrying scheme. A major advance [15] is to consider the introduction of multiple ferries and not only one. As a generalized problem, the route computation for multiple ferries is also NP-hard. Multiple cases have been considered with or without communications between the multiple ferries. Nevertheless, the use of controlled mobility in routing case is now considered for networks in which all the nodes are controlled mobility enabled and are all considered in order to make a multi-hop routing while adapting network topology to the network activity.

1.3 Network Connectivity Guarantee

The hypothesis of a complete network on which we can relocate every node on the source destination path in order to both optimize network topology and increase delivery rate is not without consequences. Network connectivity has to be maintained. In other words, it means that there must be one (or multi)-hop path between each node of the network at every moment. Two approaches can been considered. *(i)* We ensure that a moving node stays connected to the network; it means it stays connected to all of its neighbors despite movements. But this approach would be very limiting in terms of relocation possibilities. Moreover the higher the network density δ is, the more limited the movements as the moving node has to maintain connectivity to more nodes. *(ii)* We ensure that it moves in the range of at least one of its static neighbor nodes and stays connected to the network.

We now introduce some graph tools that will be later used for maintaining connectivity in wireless mobile networks: Relative Neighborhood Graph (RNG) and Connected Dominated Sets (CDS).

1.3.1 *Relative Neighborhood Graph*

Relative Neighborhood Graph (RNG) [11] is a graph reduction from a graph $G = (V, E)$ to $G' = (V, E')$ that can be computed locally. It is defined as follows. An edge (u, v) between nodes u and v exists in G' $((u, v) \in E')$ if it exists in G $((u, v) \in E)$ and if the Euclidean distance $|uv|$ is lower than or equal to the distances $|uw|$ and $|vw|$ for any other vertex $w \in V$ such that $(u, w) \in E$ and $(v, w) \in E$. This can be noted as :

$$(u, w) \in E' \equiv |uv| \geq \max_{\forall w \neq u, v \in E}[|uw|, |vw|] . \tag{1.1}$$

It reduces the average node degree to $\simeq$3 while preserving networking connectivity. A moving node which stays connected to its RNG neighbors will keep network connectivity unchanged.

1.3.2 *Connected Dominated Set (CDS)*

A Connected Dominated Set (CDS) is a connected subset of the network that covers the same area. If nodes in the CDS (*i.e., dominant*) are static, we ensure that all mobile nodes stay in transmission range of the CDS. It guarantees that their is always a path between every pair of nodes of the network. In [1] authors have proposed a fully localized algorithm. Giving a node A, $N(A)$ its set of neighbor and S the subset of $N(A)$ with higher priority (any criteria can be used, such as id, battery level, etc...) than A: $S \leftarrow N(A) - \{U \in E \mid priority(U) < priority(A)\}$, A belongs to the CDS if one of the following statement do not hold:

- S is not empty: $S \neq \{\emptyset\}$
- S is connected: $\forall A \in S, \exists B \in S\ s.a.\ A \neq B \wedge |AB < radio\_range$
- every node in $N(A)$ is in S or in range of S: $\forall B \in N(A), B \in S \cup N(S)$.

Figure 1.3 shows such a computation.

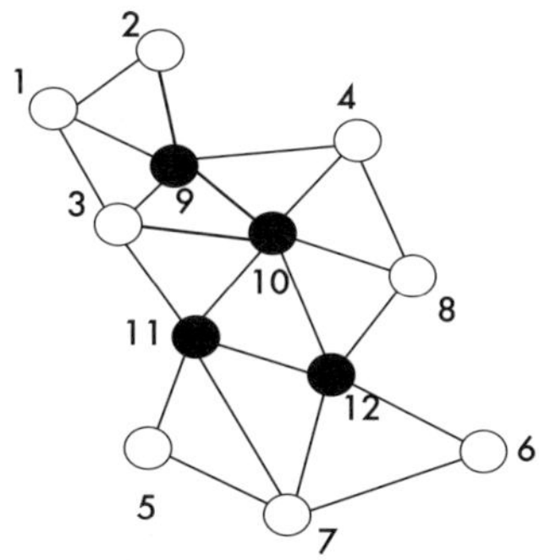

Fig. 1.3 Connected Dominating Set of a graph. Number represents node priority. Black nodes are dominant and cannot move.

1.4 Actuator Networks

Most of the existing approaches work as follows. A Route Request (RQ) is forwarded greedily to find an initial route, and then, during the Route Reply (RP), nodes on the route iteratively move to a new location on the virtual line between source and destination. All these routing protocols may not be efficient. Moving strategy in [3] may cause useless zig-zag movements as it relocates nodes on the iso-barycenter between upstream and downstream nodes. This behavioral issue is addressed in [2]. Authors propose to use virtual movements until path stabilization is reached : as long as the balance on the routing path is not achieved nodes do not move. However the approach they propose assumes that the routing path is already set.

1.4.1 *MobileCOP*

In MobileCOP [9], next hop on the path is selected following Cost-Over-Progress (COP) metrics. The paper proposes either to use the optimal computed hop count or to minimize the power consumption over the progress. Once the path is computed, nodes are moved and placed equidistantly on the straight line connecting the source to the destination, maintaining the same number of hops as in the computed path. Such moves may induce a memory overhead on nodes, since they have to store the path, and a high transmission delay. Moreover, the network may be disconnected (a node may move out of range of its neighbors). These arbitrary movements may not be efficient as when a node moves further from its upstream neighbor, the transmission delay will automatically increase.

1.4.2 *RPCM*

Routing Protocol Based on Mobility (RPCM) [10] also works with a RQ/RP mechanism but builds multiple RR paths toward the destination. This is because every neighbor of the source node will forward a RQ packet, adding its information (id and location) in the packet and forward it. The destination will consequently receive multiple RQ from different routing paths. After a predefined time, the destination selects the path in which nodes have to travel the total minimum distance to reach the evenly spaced positions on the straight line between the source-destination pair. Consequently, the destination forwards the Reply Packet (RP) in backward direction on the elected routing path. During the RP backward routing nodes relocate on the Source-Destination line such as distance between routing nodes are equal on the path.

1.4.3 *CoMNet*

A more recent and very different approach is CoMNet [7], Connectivity preservation Mobile routing protocol for actuator and sensor NETworks. CoMNet takes advantage of controlled mobility to adapt network topology to the traffic. Routing paths are constructed on the fly, there is no RR/RP mechanism : CoMNet is memoryless and does not require to store path informations in the routed messages. Moreover CoMNet is the first routing protocol relying on controlled mobility which :

- takes account to both relocation and radio-transmission costs in a COP fashion
- can guarantee network connectivity despite nodes movement using a CDS (see Section 1.3.2)
- comes in three different variants to fit the best to various environments application requirements.

The variants aim to adapt the quantity of movement or the radio transmission costs based on the application and model requirements :

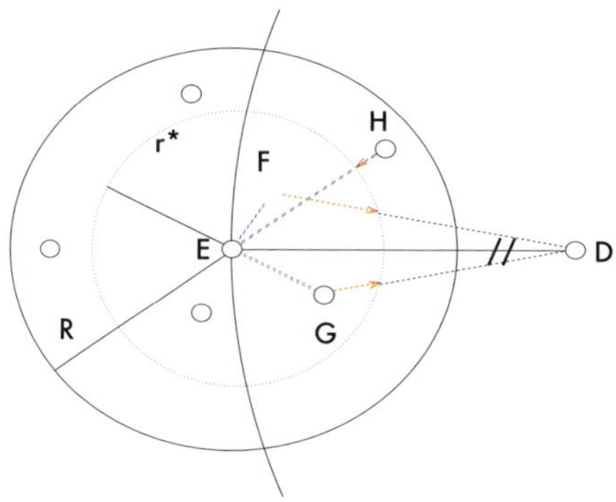

(a) CoMNet MoveDS(r)

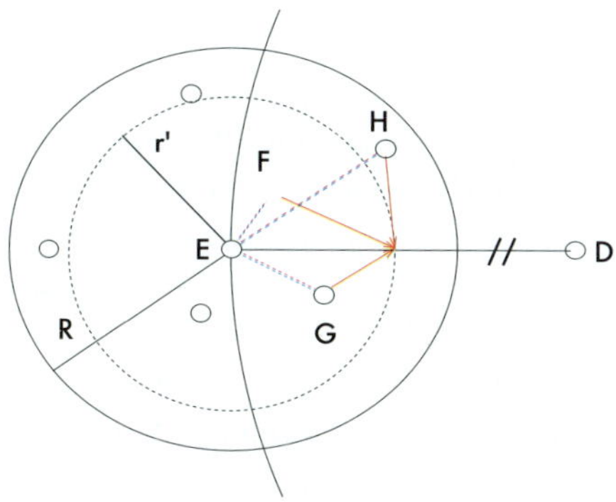

(b) CoMNet Move(r)

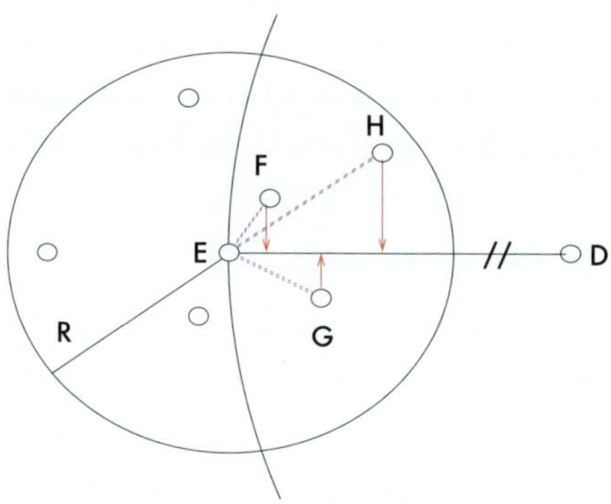

(c) CoMNet Orouting On the Move

Fig. 1.4 The three different CoMNet relocation patterns. Red arrows indicate neighbor relocation according to the relocation pattern.

- $CoMNet{-}Move_{(DSr)}$ aligns nodes on the Source Destination (SD) line with all hop lengths to be equal to the optimal transmission distance r^*. This pattern aims to temper both communication and relocation costs.
- $CoMNet - ORouting$ *on the* $Move$ aligns nodes on the (SD) line. If E selects nodes F as next forwarding node, it sends to F a *beacon* which is a relocation order with its computed new location. When F reaches its new location, F forwards it the packet to be routed. This pattern aims to limit energy consumption due to radio communications.
- $CoMNet - Move_r$ makes next hop node to be relocated on the intersection of the $C(E, r^*)$ centered in E and of radius r^* and the (ED) line, in order to limit node relocations.

Figure 1.4 illustrates the next hop selection and relocation for each of these patterns. The current node E computes the relocation pattern of each of its neighbors together with the associated cost (cost to send the message to this neighbor plus cost of the relocation of the neighbor). E then selects the ones that provides the smallest ratio cost over progress (computed as $|ED| - |xD|$ where x is the selected node after relocation). Results show that CoMNet allows great energy savings compared to MobileCop.

CoMNet has been enhanced in MobileR [5]. In MobileR, at every routing step, in order to anticipate next steps, every node computes potential next re-locations that may occur on its own neighborhood, computing in this way the cost of multi-hop relocations. This may improve the energy efficiency of the paths but requires a more complex computing at nodes. The tradeoff between what this variant brings and what it requires has thus to be analyzed regarding hardware costs and applications requirements.

1.4.4 *CoMNet-based extensions*

CoMNet has been further extended to answer issues related to path crossings and recovery. In this section, we give an overview of them.

1.4.4.1 *PAMAL*

Most available solutions, such as **CoMNet**, iteratively move each node of the Source-Destination route on a (SD) line to adapt the topology to the network activity. However, in event-based WSN, a same event often triggers multiple sensors which are close from each other, but not enough to be able to communicate between each other. In response to the event, each triggered node will emit a message toward the network sink. Each different $(S1 \rightarrow D)$... $(Sn \rightarrow D)$ route provokes a topology adaptation with regard to this specific route. But since the sources are close, different routing paths tend to intersect and share nodes. In this case, controlled mobility provokes competition and useless zig-zagS for those nodes which belong to the different $(S \rightarrow, D) \ldots (Sn \rightarrow D)$ routes. Those nodes die prematurely and multiple routing paths with them. Fig. 1.5 illustrates such a case with node E. This issue is addressed in [4] with the **PAth Merging**

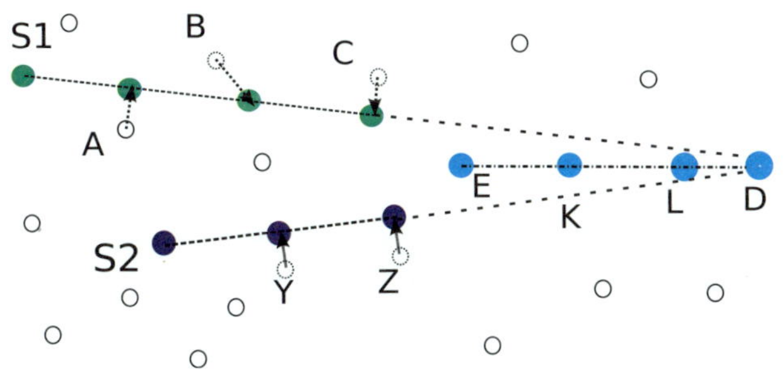

Fig. 1.5 $S1$ and $S2$ adapt the network topology on their respective routing paths. Node E belongs to both $(S1 \rightarrow D)$ and $(S2 \rightarrow D)$ and oscillates between different relocations.

Algorithm. Relying on CoMNet relocation patterns and COP approach, **PAMAL** detects path crossings and handles them in a purely localized way. Moreover, those intersection nodes which belong to multiple paths, are used as temporary destinations in the relocation pattern, making crossing paths physically merge along time. PAMAL is illustrated in Fig. 1.6 which is a zoom on the E node of Fig. 1.5. E has detected the path crossing. First, it stops oscillating between routing paths an goes to the iso-barycenter of previously asked relocations. In a second step it advertises its neighbors to use it as a temporary destination. Hence nodes B, C, Y and Z computations and relocations evolved making $(S1 \rightarrow D)$ and $(S2 \rightarrow D)$ paths merge.

1.4.4.2 *GRR*

Greedy Routing Recovery (GRR) [6] is a routing protocol uses controlled mobility in order to bypass topology holes where density is not sufficient to allow greedy forwarding up to the destination. GRR combines the greedy

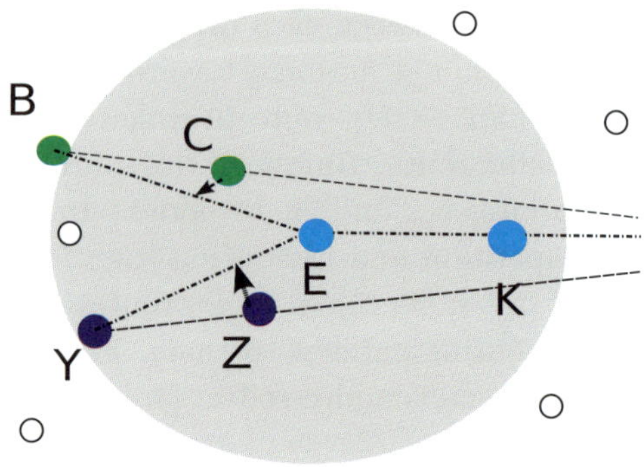

Fig. 1.6 Node E advertises its neighbors to consider its own location - instead of the one of D- for relocation patterns. ($S \rightarrow D$) and ($S2 \rightarrow D$) paths merge.

forwarding of CoMNet with a light recovery mechanism, allowing less calls to the costly recovery process. In its greedy forwarding step, GRR routes packet and relocates nodes in a CoMNet fashion. Routing nodes are moved on the source-destination line in order to make energy savings. Nevertheless, when the greedy routing fails, GRR switches to light recovery step.

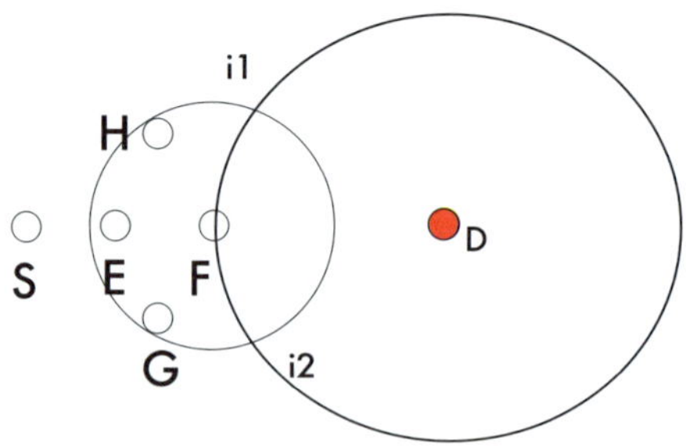

Fig. 1.7 Routing from S to D fails on node F. Greedy routing could go on if one of F's neighbors was on $i1$ or $i2$.

In light recovery, the relocation pattern is completely different from CoMNet one. Plain CoMNet routing fails because the forwarding node has no neighbor closer to the destination than itself. In Fig. 1.7 routing from S to D fails on F. GRR light recovery will make forwarding node F computes the $i1$ and $i2$ locations and the cost for every of its neighbors (H, E and G) to move up to it. Fselects the node which minimizes this cost as next forwarder. By doing so, GRR makes possible the bypass of the routing hole and restores greedy routing on the next routing for one hop more.

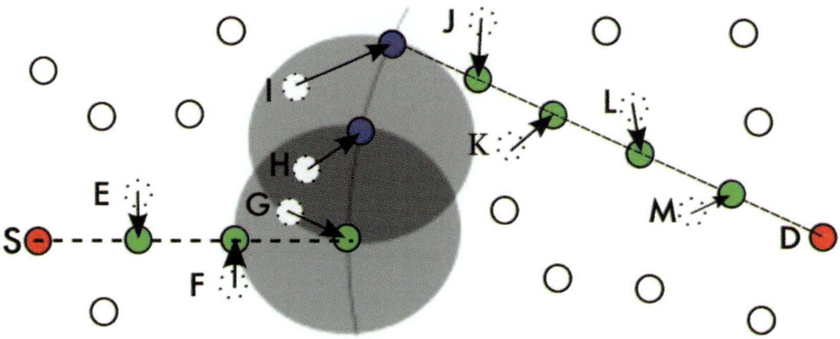

Fig. 1.8 First GRR routing from S to D. Greedy routing part of the path is in green, while the light recovery is in blue.

Figure 1.8 illustrates a hole bypassing using GRR routing. From S to G, greedy forwarding is performed (in green) using CoMNet relocation patterns. But node G has no neighbor closer to D than itself. It switches to light recovery and relocates H. Greedy routing is still impossible on relocated H, light recovery goes on up to node I. Finally, the greedy routing is possible from relocated I up to D. On the next routing from S to D, the whole forwarding will be performed in a greedy manner.

Simulations demonstrate that GRR successfully bypasses topology holes in more than 50% of network topologies avoiding calling to expensive recovery steps and reducing energy consumption while preserving network connectivity.

1.5 Conclusion

Node mobility has long been considered as a hazard for geographic routing, causing a degradation of performance or even persistent routing failures. Therefore, little work has been done in routing under the assumption of mobile nodes. The introduction of *controlled* mobility has made possible the emergence of two new routing paradigms. The first one is Message Ferrying. It supposes that one or a set of dedicated controlled mobility enabled nodes called ferries will be completely dedicated for routing. Ferries travel across the network following usually predefined trajectories. They gather messages, store them, and then they carry the messages over their route until they forward or deliver them. Such an approach makes possible routing in partitioned networks or saves huge amounts of energy since the route design is supposed to be optimized with regard to the network activity. The second approach relies on the assumption that all nodes are controlled mobility enabled. Consequently routing protocols can adapt the network topology to the network activity in a very sharp, dynamic and local manner. They only move nodes on the routing paths in a efficient manner. Advanced routing topics are now being considered such as delivery guarantee (GRR) or quality of service making routing with controlled mobility an hot research topic.

References

[1] Carle, J. and Simplot-Ryl, D. (2004). Energy-efficient area monitoring for sensor networks, *Computer* **37**, 2, pp. 40–46, doi:10.1109/MC.2004.1266294.

[2] Costanzo, C., Loscrì, V. and Natalizio, E. (2009). Distributed virtual-movement scheme for improving energy efficiency in wireless sensor networks, in *Proceedings of the 12th ACM international conference on Modeling, analysis and simulation of wireless and mobile systems*, MSWiM '09 (ACM, New York, NY, USA), ISBN 978-1-60558-616-8, pp. 297–304, doi: 10.1145/1641804.1641855.

[3] Goldenberg, D., Lin, J., Morse, A. S., Rosen, B. E. and Yang, Y. R. (2004). Towards mobility as a network control primitive, in *Proceedings of the 5th ACM international symposium on Mobile ad hoc networking and computing*, MobiHoc 2004 (ACM, New York, NY, USA), ISBN 1-58113-849-0, pp. 163–174, doi:10.1145/989459.989481.

[4] Gouvy, N., E., H., Zorbas, D. and Mitton, N. (2013a). Energy Efficient Multi-Flow Routing in Mobile Sensor Networks. in Proceedings of the IEEE Wireless Communications and Networking Conference, WCNC 2013 (Shanghai, China), URL `http://hal.inria.fr/hal-00772217`.

[5] Gouvy, N., Elhafsi, E. H., Mitton, N. and Simplot-Ryl, D. (2012). M. energy consumption through mobility with connectivity preservation in sensor networks, *IJPEDS* **27**, 6, pp. 521–540.

[6] Gouvy, N., Mitton, N. and Zheng, J. (2013b). Greedy Routing Recovery Using Controlled Mobility in Wireless Sensor Networks. in Procedings of the 12th International Conference on Ad Hoc Networks and Wireless, AdHoc-Now 2013 (Worclaw, Poland).

[7] Hamouda, E., Mitton, N. and Simplot-Ryl, D. (2011). Energy Efficient Mobile Routing in Actuator and Sensor Networks with Connectivity preservation. in Proceedings of the *10th International Conference on Ad Hoc Networks and Wireless, AdHocNow 2011* (Padderborn, Germany), pp. 15–28.

[8] Li, Q. and Rus, D. (2000). Sending messages to mobile users in disconnected ad-hoc wireless networks, in *Proceeding of the 6th annual international*

conference on Mobile computing and networking, MobiCom '00 (ACM, New York, NY, USA), ISBN 1-58113-197-6, pp. 44–55, doi:10.1145/345910.345918.

[9] Liu, H., Nayak, A. and Stojmenovic, I. (2007). Localized mobility control routing in robotic sensor wireless networks, in H. Zhang, S. Olariu, J. Cao and D. Johnson (eds.), *Mobile Ad-Hoc and Sensor Networks*, *Lecture Notes in Computer Science*, Vol. 4864 (Springer Berlin Heidelberg), ISBN 978-3-540-77023-7, pp. 19–31, doi:10.1007/978-3-540-77024-4_4.

[10] Loscrí, V., Natalizio, E. and Costanzo, C. (2010). Simulations of the impact of controlled mobility for routing protocols, *EURASIP J. Wirel. Commun. Netw.* **2010**, pp. 7:1–7:12, doi:10.1155/2010/315381.

[11] Toussaint, G. T. (1980). The relative neighbourhood graph of a finite planar set, *Pattern Recognition* **12**, pp. 261–268.

[12] Wang, W., Srinivasan, V. and Chua, K.-C. (2008). Extending the lifetime of wireless sensor networks through mobile relays, *IEEE/ACM Transactions on Networking* **16**, 5, pp. 1108–1120, doi:10.1109/TNET.2007.906663.

[13] Zhao, W. and Ammar, M. (2003). Message ferrying: proactive routing in highly-partitioned wireless ad hoc networks, in Proceedings of the 9th IEEE Workshop on Future Trends of Distributed Computing Systems, FTDC 2003 (San Juan, Puerto Rico), pp. 308–314, doi:10.1109/FTDCS.2003.1204352.

[14] Zhao, W., Ammar, M. and Zegura, E. (2004). A message ferrying approach for data delivery in sparse mobile ad hoc networks, in *Proceedings of the 5th ACM international symposium on Mobile ad hoc networking and computing*, MobiHoc '04 (ACM, New York, NY, USA), ISBN 1-58113-849-0, pp. 187–198, doi:10.1145/989459.989483.

[15] Zhao, W., Ammar, M. and Zegura, E. (2005). Controlling the mobility of multiple data transport ferries in a delay-tolerant network, in Proceedings of the 24th Annual Joint Conference of the IEEE Computer and Communications Societies. INFOCOM 2005 (miami, Florida), Vol. 2, pp. 1407–1418 vol. 2, doi:10.1109/INFCOM.2005.1498365.

Chapter 2

Accelerated Random Walks for Efficient Data Collection in Mobile Sensor Networks

Angelopoulos Constantinos Marios and Sotiris Nikoletseas

University of Patras and CTI & Press "Diophantus", Greece

Abstract. *Random walks in Wireless Sensor Networks can serve as fully local, very simple strategies for sink motion that lift from sensors the computational burden of propagating collected data via complex routing algorithms. In such schemes, sensors maintain a passive role waiting for the sink to visit them and collect their data. Therefore, energy dissipation is significantly reduced at the expense of increasing the latency of data delivery. To achieve satisfactory energy-latency trade-offs, the sink walks can be made adaptive, depending on critical network parameters (density, history of past visits in each network region) as well as characteristic attributes of the network topology. In this chapter, we study known random walks on two well established network models; the Grid and the Random Geometric Graphs. Towards better balances of memory/performance, we propose four new random walks that we experimentally evaluate in terms of several performance metrics that capture important differences in the network traversal process. The simulation findings demonstrate that even with limited memory usage, random walks can prove themselves very efficient when carefully designed.*

2.1 Introduction

Wireless Sensor Networks are envisioned as large ad-hoc collections of very small autonomous devices, that can sense environmental conditions in their immediate surroundings while having limited processing, communication capabilities and energy reserves (see e.g., [12] for a comprehensive coverage of the fundamental algorithms and protocols for wireless sensor networks). The collected sensory data is usually disseminated to a static control center (called data sink) in the network, using node to node *multi-hop* data propagation. Such settings have increased implementation complexity and sensor devices consume significant amounts of energy. This is due to the fact that multi-hop data propagation protocols in sensor networks, where the sink is static, leads the sensor devices to consume significant amounts of energy in operations other than sensing (e.g., inter-node communication for synchronization purposes, exploratory messages, etc). Moreover, the data exchanged are used as input to several complex distributed algorithms and protocols (such as localization) run by the sensor devices. Even though those algorithms are designed so as to consume as less resources as possible, the problem still exists. Furthermore, in the area around the control center, nodes need to heavily relay the data from the entire network, thus a bottleneck of increased energy consumption emerges and failures due to strained energy resources of these nodes lead to an early disconnected and dysfunctional network (see e.g., [10]).

Towards a more balanced and energy-efficient method of data collection, sink mobility can be used. The sink is lifting the burden of inter-node coordination by traversing the network itself. A mobile sink may also substitute connectivity as it is capable of bypassing obstacles and reaching disconnected components of the network, while sensor motes maintain a rather passive role, in terms of data propagation, by simply waiting to contact the sink. Once contact is established, they deliver data via cheap (energy wise) one-hop transmissions. However, many apparent difficulties arise as well since traversing the network area in a timely and efficient way is critical. Failure to visit some areas of the network will result in data losses, while infrequently visiting some regions will introduce high delivery delays and memory overflows.

Apparently, in such data collection schemes, new critical issues emerge regarding the pattern of movement to be adopted by the sink. The network has to be traversed in a timely and efficient manner. Also, the mobility pattern of the sink should guarantee that the entire network area, or at least the vast part of it, will eventually be visited.

In the light of the above discussion, random walks in wireless sensor networks can serve as fully local, very simple, distributed strategies for sink motion that significantly reduce energy dissipation and also probabilistically guarantee that eventually the entire network area will be covered. Several real life applications exploit sink mobility for more efficient data collection, e.g., robots that move inside a building that needs to be evacuated or foresters moving through a forest where sensors have been deployed.

In this chapter, we study random walks as mobile sink strategies in Wireless Sensor Networks over two characteristic network models. At first, we abstract the network via a Grid that is virtually overlaid over the network area; a mobile sink traverses the area by moving on the edges of the graph. By restricting the random walk to an overlaid graph and by taking advantage of its regular structure, we are able to perform some optimizations and gain an insight to several aspects of the problem. On this model we further investigate four already known walks particularly suitable for data collection in WSNs where sensors are static. Towards improved trade-offs, we then propose three new walks that require small memory of constant size. Following, we choose to model wireless sensor networks via Random Geometric Graphs (RGG), that better capture certain relevant characteristics of real WSNs such as link existence dependencies of neighboring nodes due to geometric proximity. On this model we revisit several random walks already studied in the Grid and we propose a fourth new random walk.

Finally, we experimentally evaluate the performance of each walk both in terms of well established metrics (cover time, approximate cover time, visit overlap statistics) and novel ones (proximity variation and proximity cover time) that capture the differences on how each walk traverses the network area. Our goal is mainly to accelerate the random walk coverage of the network via reducing the node overlaps. Also, to improve other (network related) features of the walks, such as how fast they get close to the network nodes of different network subregions. Some of the walks discussed in this chapter, as well as corresponding preliminary results have been first presented in [1] and [2].

2.2 State of the Art

Random Walks have been extensively studied in the past decades in the context of several disciplines. However, despite their numerous applications in ad hoc, VANETs (see e.g., [13]) and overlay networks, not much research has been done on how they can be particularly applied in sensor networks with mobile entities, in a way that addresses the peculiarities of such networks (such as severe computing and communication constraints, the small memory and constrained battery, as well as the time-criticality of important applications).

A comparison of different random walk strategies for ad hoc networks is performed in [6]. The authors investigate the effectiveness of each strategy in terms of the expected hop count and the occurrence of deadlocks. In this chapter, we present a new random walk strategy and compare two variations of it with two known walks. We study the performance of the strategies in terms of cover time, approximate cover time, proximity variation and proximity cover time.

In [15] the problem of data gathering in a large-scale wireless sensor network with static nodes and a mobile patrol node is formulated as a classical random walk on a random geometric graph. The authors derive analytical bounds for the performance of the random walk in terms of node coverage. In order to improve this performance they propose an algorithm to constrain the random walk using the available side information, such as the awareness of previously visited sites. The walks in this chapter also use some information (just the information of geometric distance) in order to speed up the network traversal and our new walks are different.

In [20] authors investigate the *Random Waypoint Model*, which is widely used in the simulation studies of mobile ad hoc networks and show that it fails to provide a steady state in that the average nodal speed consistently decreases over time; therefore it should not be directly used for simulation. They also propose a simple fix of the problem and discuss a few alternatives. In our study, the moving element of the network (the sink) is visiting the nodes in a sequential way. Therefore, we abstract its motion by considering that it is moving on the edges of a Random Geometric Graph, that models a wireless sensor network.

Nice research has also been conducted for deterministic sink mobility, such as in [16, 17] where authors address the problem of maximizing the network lifetime. In [16], authors suggest that the base station can be mobile and conclude that the best mobility strategy consists in following the periphery of the network (assuming that the sensors are deployed within a circle). In [17], authors investigate the approach that makes use of a mobile sink for balancing the traffic load and in turn improving network lifetime. They engineer a routing protocol that effectively supports sink mobility and through intensive simulations, they prove the feasibility of their mobile sink approach.

In [11] the authors study the problem of data collection or in a sensor network using a mobile sink. Towards improved latency, they propose a biased, adaptive sink mobility scheme. The sink moves probabilistically, following a biased random walk that favors less visited areas in order to cover the network area faster, while it adaptively stops more time in network regions that tend to produce more data. The proposed random walk strategies in this chapter instead use only the information of their previous positions in order to decide their next move, in contrast to [11] which assumes memory of $O(n)$, where n is the number of nodes of the graph. However, despite the very small memory that we here use, we succeed to cover the network much faster than a simple random walk and in some respect (approximate cover time, proximity variation), we even outperform the powerful biased walk of [11].

In [14], the authors investigate the theoretical aspects of the uneven energy depletion phenomenon around a sink,and address the problem of energy-efficient data gathering by mobile sinks. The authors also present a taxonomy and a comprehensive survey of the state of the art on the topic.

Random walks have also been used to model different types of interactions in mobile computing scenarios, such as in [8, 18] and to analyze communications in mobile computing, e.g., in [4, 7] the authors propose protocols which exploit the coordinated, random motion of a small part of the network; they also propose a methodology for the analysis of the expected behavior of protocols for such networks, based on the assumption that mobile hosts (whose motion is not guided by the protocol) conduct concurrent random walks in their motion space. In contrast, the walks in this chapter are not blind but instead adapt to the progress of the data collection process.

2.3 The Network Models

Sensor networks comprise of a vast number of ultra-small homogeneous sensor devices (which we also refer to as sensors), whose purpose is to monitor local environmental conditions. Each sensor is a fully-autonomous computing and communication device, characterized mainly by its available power supply (battery), its transmission range r, the energy cost of data transmission and the (limited) processing and memory capabilities. Sensors (in our models here) do not move. The positions of sensors within the network area are random and follow a uniform distribution. We focus on data collection methods, so we assume that initially all sensors have some data to deliver to the sink. For clarity, we also assume that no data is generated during the network traversal. That is, if a node is called "visited", then it has no data to send to the sink.

Sensor devices are equipped with hardware monitors that measure environmental conditions of interest, including light, pressure, humidity, temperature etc. Each device has a *broadcast* (digital radio) *beacon mode* of fixed transmission range $\mathcal{R}$, and is powered by a battery. Also a sensor is equipped with a *general purpose storage memory (e.g., FLASH) of size* $\mathcal{C}$.

There is a special node within the network region, which we call the sink S, that represents a control center where data should be collected. Here, we assume that the sink is *mobile*. The sink is not resource constrained i.e., it is assumed to be powerful in terms of computing, memory and energy supplies.

We study random walks as mobile sink strategies in Wireless Sensor Networks over two characteristic network models; namely the *Grid* and the *Random Geometric Graphs*. In the first case, the network area A is a flat square region of size $D \times D$. During the network initialization, a graph formation phase is executed by the sink. The network area is partitioned in $j \times j$ equal square regions, called cells. The center of each cell is considered as a vertex in a graph that is connected with non-directional edges only to the four vertices corresponding to adjacent cells. Thus, a virtual lattice graph $G_o = G(V, E)$ is created which is overlaid over the network area. We set $j = \lceil D/\sqrt{2}\mathcal{R} \rceil$, thus when the sink is located at the center of a cell, it can communicate with every sensor node within the cell area. By reducing the walk to an overlay graph we can perform some optimizations more easily by taking advantage of its regular structure. Also, the mobility schemes can be deployed in areas of arbitrary topologies as long as they can be abstracted by an overlay graph.

In the second case, we consider that the random uniform placement of the sensors inside the network area is abstracted by a Random Geometric Graph. Random Geometric Graphs are formed by n vertices that are placed uniformly at random in the $[0,1]^2$ square. An edge (u, v) exists iff the Euclidean distance of vertices u and v is at most r, where r corresponds to the wireless communication radius r of the sensors. This holds assuming a disc radio model; two sensors can communicate with each other iff each one lies inside the communication range of the other. Random Geometric Graphs also have an important nice property: unlike other random graphs, like $G_{n,p}$, edges are not statistically independent of each other. That is, the existence of an edge (u, v) is not independent of the existence of edges (u, w) and (w, v). This property makes RGG quite a realistic model for wireless sensor networks that captures to a great extent the communication structure of real WSNs (at least their spatial aspects).

More strictly, consider an area $\mathcal{A} \subset \mathbb{R}^2$ in two dimensional space. An instance of the *random geometric graphs model* $\mathcal{G}(\mathcal{X}_n; r)$ is constructed as follows: select n points $\mathcal{X}_n$ uniformly at random in $\mathcal{A}$. The set $V = \mathcal{X}_n$ is the set of vertices of the graph and we connect two vertices if their Euclidean distance is at most r. For any vertex $v \in V$ we will denote by $N(v)$ the set of neighbors of v and by $\deg(v) = |N(v)|$ its degree. Furthermore, we will denote by $\|u - v\|$ the Euclidean distance between the points corresponding to vertices v, u.

In [9, 19] it is shown that the connectivity threshold for $\mathcal{G}(\mathcal{X}_n; r)$ is $r_c = \sqrt{\frac{\ln n}{\pi n}}$. In this chapter we will consider random instances of $\mathcal{G}(\mathcal{X}_n; r)$ of varying density, by selecting $r = \sqrt{\frac{c \ln n}{\pi n}}$, for different values of $c > 1$, which guarantees that the produced random instance is connected with high probability.

Besides the information about the set of neighbors of each vertex $v \in V$, an instance $G(\mathcal{X}_n; r)$ of the random geometric graphs model also contains extra information about the exact Euclidean distance between v and any of its neighbors. We used this information in order to define a new random walk, namely the *γ-stretched random walk* on $G(\mathcal{X}_n; r)$, which is described below. This random walk aims to accelerate the data collection process while keeping the memory requirements restricted. The basic idea is that "more distant" neighbors of the current vertex (state of the walk) are favored in a probabilistic manner. Therefore, our new walk is in fact a *biased transitions* random walk (see e.g., [3] for a nice discussion of several types of random walks based on the assumptions they made).

2.4 The Sink Mobility Random Walk Protocols

2.4.1 *Classic Random Walks*

2.4.1.1 *Blind Random Walk*

This is the usual random walk model. The Blind Random Walk is the simplest of all possible sink mobility patterns, since the next move of the sink is stochastically independent to the previous ones. Furthermore, given that the current vertex is $v \in V$, the probability of moving to any neighboring vertex $u \in N(v)$ is $p_{v,u} = \frac{1}{\deg(v)}$. This method is very robust, since it probabilistically guarantees that eventually all network regions and nodes will be visited and thus all data will be collected given that the network is connected. However, in some network structures it may become inefficient, mostly with respect to high latency, since the sink uses no memory of the past movements in order to select the next one and thus overlaps (i.e., visits to already visited vertices) occur.

2.4.1.2 *Random Walk with Memory*

The performance of the Blind Random Walk can be improved using some memory of past visits. In Random Walk with Memory K, the sink maintains a first-in-first-out (FIFO) list $\mathcal{M}$ which contains the last K nodes visited during the random walk, i.e $M = \{c_1, c_2, ..., c_K\}$. The next hop is chosen uniformly at random among the neighbors of the node that are not in the memory list $\mathcal{M}$. The use of memory eliminates the possibility of short loops in random walks. Setting $K = 0$, we simply get the blind random walk. In this study, the random walk with memory 1 will be used for comparison purposes. Note that this walk has stronger memory requirements and overhead than the blind random walk.

2.4.1.3 *Biased Random Walk*

We investigate the Biased Random Walk only on the Grid model. In this walk, the sink associates a counter c_u for every vertex u; initially $c_u = 0$ $\forall u \in V$. When the mobile sink visits cell u, it increases the associated counter c_u by 1. Thus, the frequency of visits of each area can be estimated and maintained by the sink. The selection of the next area to visit is done in a biased random manner depending on this variable. If the mobile element is currently in cell u of degree $\deg(u)$, then we define

$$c_{neigh}(u) = \sum_{v} c_v,$$

for all $v : (u, v) \in E$, i.e., $c_{neigh}(u)$ is the total number of past visits in the neighborhood of vertex u. Then the probability $p(f)_v$ of visiting an adjacent region v is taken as

$$p(f)_v = \frac{1 - c_v/c_{neigh}(u)}{\deg(u) - 1}$$

when $c_{neigh} \neq 0$. When $c_{neigh} = 0$ we have $p(f)_v = 1/\deg(u)$. Thus, less frequently visited regions are favored when the sink is located at a nearby region. So, with the use of the entire history of visits in each cell, the trajectory of the sink is improved and the time needed for traversing the whole network reduced, since overlaps are probabilistically discouraged. However, this walk requires a memory of size $O(n)$, where n is the number of cells.

2.4.2 *Adaptive Random Walks*

2.4.2.1 *Random Walk with Inertia*

In the heuristic Random Walk with Inertia *on the Grid model*, the sink assigns a probability to each of the four directions, North, South, East and West. At the beginning the probability distribution is uniform, i.e., the sink has the same probability to move towards any of the four directions. Afterwards, the probability distribution at each step of the walk changes adaptively to the nodes' discovery following the principle of *reinforcing the direction where newly discovered sensors were found and weakening the direction where already visited sensors have been located.* For instance, if the sink discovers new sensors after having gone West, then the probability of choosing the same direction for the next step is increased by δ, while the rest probabilities are reduced by $\frac{\delta}{3}$ each, as long as they remain non-negative. δ is a constant probability; its specific value is optimized in light of the detailed experimental findings. We considered several $\delta \in [0.1, 0.5]$ and for each one compared the experimental results. The best performance resulted when $\delta = 0.2$. The reverse process of weakening the previously chosen direction and reinforcing the other three is followed when the sink reaches a cell that has already been visited. To summarize, the probability of following the same direction (p_c: current direction probability) at time t is:

$$p_c^{t+1} = \begin{cases} p_c^t + \delta & \text{, if new nodes discovered} \\ p_c^t - \delta & \text{, if no new nodes discovered} \end{cases}$$

while each one of the probabilities towards the other three directions (p_r: rest direction probabilities) are:

$$p_r^{t+1} = \begin{cases} p_r^t - \frac{\delta}{3} & \text{, if new nodes discovered} \\ p_r^t + \frac{\delta}{3} & \text{, if no new nodes discovered.} \end{cases}$$

Already visited cells are distinguished from newly discovered ones by simply checking whether the sensors of the cell have data to send or not, i.e., we first visit and then check if visited or not (this is in contrast to the Biased Random Walk that assumes knowledge of the visit history).

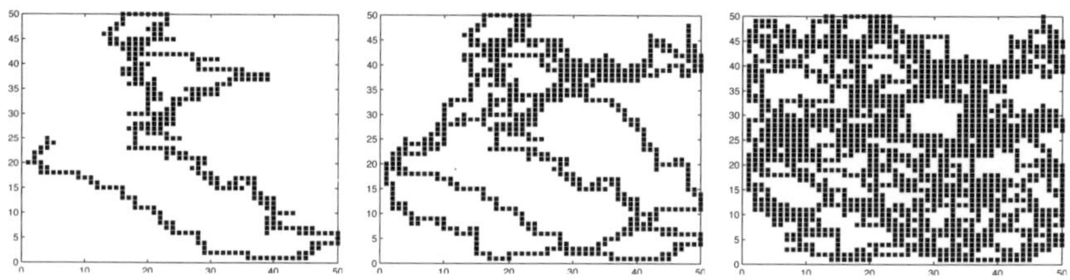

Fig. 2.1 An example of network traversal following Random Walk with Inertia in a 50×50 network area (snapshots after: 500, 1000 and $2 * 10^4$ hops).

On the Random Geometric Graph model (RGG), we adapt the walk as following. Assume an instance $G(\mathcal{X}_n; r)$ of the RGG model. Let v denote the current state of the walk, i.e., the current vertex. Let S^v denote the set of vertices that are 1-hop neighbors of v. Finally, let $v_{previous}$ denote the previous state of the walk. At the beginning of the walk given its current state, the next hop is chosen uniformly at random among the neighboring nodes and $v_{previous}$ is not defined. From the second hop and so forth the next state of the walk is chosen among the vertices of the set $S^v / v_{previous}$ according to the following probability distribution function:

$$P\{next\_state == i\} = \begin{cases} \frac{\phi_i^v}{\sum_{j \in S^v} \phi_j^v} & \text{if } i \in S^v \\ 0 & \text{otherwise} \end{cases}$$

where ϕ_j^v is the angle defined by vertices v, $v_{previous}$ and j and $0 \leq \phi_j^v \leq \pi$.

We note that this walk has light-weight requirements. It assumes zero knowledge of the network and is relatively simple with low computational complexity. Furthermore, it requires a small, constant sized memory since the next step of the walk depends solely on the previous one. The sink tends to keep the same direction as long as it discovers new nodes, while it changes its direction when it encounters already visited ones. This way, the sink makes long paths and traverses many different sub-regions of the network area very quickly, thus avoiding early overlaps. However, after most of the network area has been covered, there exist small unvisited sub-regions that are hard to find (see Fig. 2.1). At this mature stage of the network traversal, the motion of the sink tends to be very similar to the Blind Random Walk, since already visited cells lie very probably, towards every direction lie already visited cells. The fact that many different sub-regions are visited very soon makes this walk suitable for time critical applications, such as reactive event detection.

2.4.2.2 *Explore-and-Go Random Walk*

We investigate the Explore-and-Go Random Walk only on the Grid model. In this walk heuristic, the movement of the sink consists in two types of motion: a) moving on a straight line and b) arbitrarily changing direction. The sink on each step chooses to follow one of these two motions, based on a constant bias factor β. For example if the bias factor is $\beta = 0.9$, then with 90% probability the sink will move towards the same direction, during the next step, while with 10% probability it will arbitrarily change its direction. There are two different values that are assigned to the bias factor during the walk, a low one and a high one. For as long as the sink keeps discovering new nodes, the bias factor β is assigned the low value. On the contrary, when the sink reaches an area that has already been visited, the bias factor is assigned the high one. Intuitively, this means that as long as there are undiscovered nodes on the nearby sub-regions of the network, the sink tends to make a Brownian-like motion, by frequently changing its direction in an arbitrary way, until all this area is covered. When no new sensors are discovered, it tends to perform a straight-line walk in order to reach a different and possibly unvisited area. The specific values of factor β were optimized in light of the detailed experimental findings, during the algorithm engineering phase. The values that give the best results are as following (F_{motion} gives the next mobility pattern chosen):

$$F_{motion} = \begin{cases} \textit{move straight, with probability } \beta \\ \\ \textit{change direction, with probability } 1-\beta \end{cases}$$

$$\text{where, } \beta = \begin{cases} 0.1 & \text{, when new nodes were discovered} \\ \\ 0.9 & \text{, when no new nodes were discovered} \end{cases}$$

We note, that this walk has light-weight requirements, as well. It also assumes zero knowledge of the network area and is relatively simple, with low computational complexity. Finally, it requires a small, constant sized memory as the next step of the walk depends only on the current position of the sink. Following this walk allows the sink to systematically cover the network area and therefore greatly reduces overlaps by avoiding to leave small unvisited sub-regions that are difficult to find during the final phases of the network traversal, when most of the network area is covered (see Fig. 2.2). On the other hand there are sub-regions that take too long for the sink to visit. Therefore, this walk is suitable for delay-tolerant applications and scenarios, which however, may require consistent data.

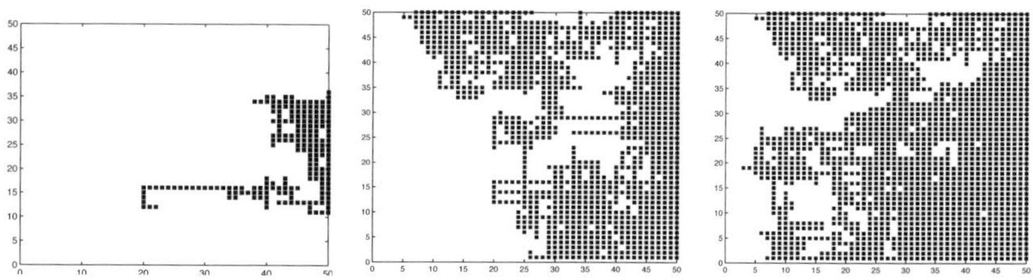

Fig. 2.2 An example of network traversal following Explore-and-Go Random Walk in a 50×50 network area. (snapshots after 500, 10^4 and $2 * 10^4$ hops.)

2.4.2.3 *Curly Random Walk*

We investigate the Curly Random Walk only on the Grid model. The main concept behind this walk is to start by visiting a confined sub-region at the centre of the network and gradually allowing the sink to perform a motion of higher degree of freedom, thus eventually covering the entire network area (see Fig. 2.3). We try to achieve this by initially having the sink perform frequent narrow left turns which gradually get wider. This way the central sub-regions of the network are first visited and the rest areas, that lie towards the boundaries of the network, are consecutively visited.

The aforementioned type of motion can be modeled as a series of successive straight and left-turn moves. Let S denote a straight move towards the same direction as the last move and L denote a left turn. A series with the desired properties, that is to initially perform narrow left turns which gradually get wider, could be of the following form:

$$SLSSLSSSLSSSSL...$$

The probability distribution of left turns in this series is the geometric distribution, with probability mass function of i straight moves followed by 1 left turn is $P^i = (1 - p_L)^i p_L$, where p_L denotes the probability of a left turn and i the number of successive straight moves before the next left turn. We call these i straight moves plus the next left turn, $phase_i$.

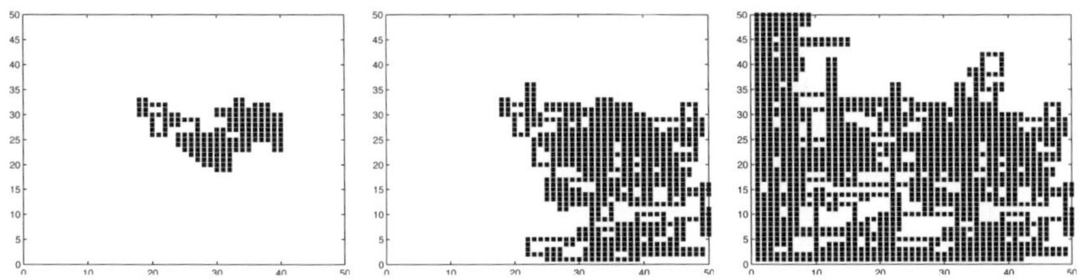

Fig. 2.3 An example of network traversal following Curly Random Walk in a 50×50 network area. (snapshots after 500, $2 * 10^3$ and $2 * 10^4$ hops.)

Note, that during $phase_i$,

$$p_L^i = \frac{1}{i+1}$$

where p_L^i is the probability of a left turn. This is chosen according to the geometric distribution, towards an i+1 expected number of trials until the first left turn.

However, since each move is stochastically independent of all previous moves, the geometric distribution does not guarantee that the desired sequence of moves will be generated. For instance, sequence SSL has the same probability to occur as sequences SLS and LSS. In order to maximize the likelihood of the desired sequence to occur but also to minimize the gaps among visited nodes in each network sub-region, we allow the sink to remain in $phase_i$ for a certain amount of time. Through experimental fine tuning we identify the best possible choice in the particular setting and allow the sink not to shift to the next phase, until 30 left turns have occurred.

We call this walk Curly Random Walk because of the fact that it converges to a walk consisting of straight lines with curls, due to successive left turns. It requires very small memory as the only information the sink remembers is the number of left turns it has taken. No information about where or when these turns happened is kept. Here we have to note that the sink can present similar behavior by increasing i at regular time intervals, which can a priori be decided (i.e., each $phase_i$ lasts for 5 minutes).

2.4.2.4 *The γ-Stretched Random Walk*

We investigate the γ-Stretched Random Walk only on the Random Geometric Graphs model. Let $G(\mathcal{X}_n; r)$ be a random instance of the random geometric graphs model with vertex set $V = \mathcal{X}_n$, where $|V| = n$. Consider a particle moving on the vertices of the graph $G(\mathcal{X}_n; r)$. Given that it occupies a specific vertex $v \in V$ at time $t \geq 0$, it decides where to move at time $t+1$ by choosing a vertex $u \in V$ with probability

$$p_{u,v} = \begin{cases} \frac{\|u-v\|^\gamma}{\sum_{w \in N(v)} \|w-v\|^\gamma} & \text{if } u \in N(v) \\ 0 & \text{otherwise.} \end{cases} \tag{2.1}$$

The Markov chain describing the above process will be called *γ-stretched random walk on $G(\mathcal{X}_n; r)$* and will be denoted by $\mathcal{W}^{(\gamma)}_{G(\mathcal{X}_n;r)}$. Furthermore, we will denote by $\mathcal{W}^\gamma_v(t)$ the state of the walk that begins at v at time t. More formally, the state space of the γ-stretched random walk is the set of vertices $V = \mathcal{X}_n$ of the graph and its transition probability matrix is given by $P = [p_{u,v}]_{u,v \in V}$.

This new random walk has minimum memory requirements, since every step is decided only by using information of the current state. Notice also that the larger γ is, the more distant neighbors are favored over neighbors that are close by (Fig. 2.4). In the simulation, we will often set $\gamma = \frac{r}{r_c}$, so that the bias to visit distant neighbors is stronger as the density of $G(\mathcal{X}_n; r)$ increases. The special case $\gamma = 1$ will be referred to simply as stretched random walk.

2.5 Performance Metrics

In the following, let $\mathcal{W}$ denote any random walk model defined either on the Grid or on a random instance $G(\mathcal{X}_n; r)$ of the Random Geometric Graphs model.

2.5.1 *Cover time and approximate cover time*

Consider the random walk $\mathcal{W}_v$ that begins on vertex v of a random instance of the random geometric graphs model $G(\mathcal{X}_n; r)$. Define $T_v \stackrel{def}{=} \inf\{t | \forall u \in V, \exists t' \leq t : \mathcal{W}_v(t') = u\}$ to be the time needed until $\mathcal{W}_v$ has visited all the vertices in the graph. The *cover time* C of $G(\mathcal{X}_n; r)$ is defined as $C = \max_{v \in V} \mathbb{E}[T_v]$, where $\mathbb{E}$ denotes the expected value of the random variable

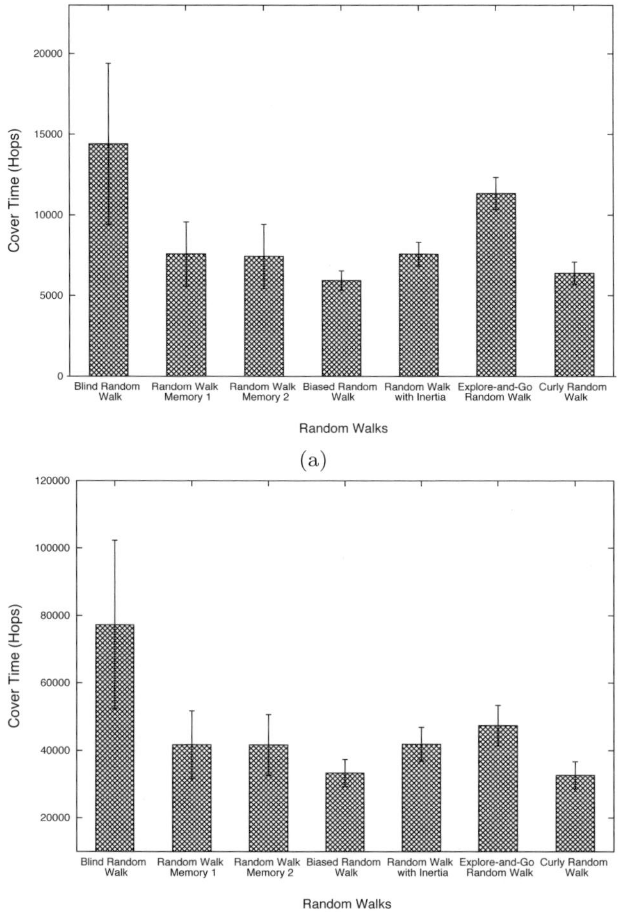

(a)

(b)

Fig. 2.4 Cover Time (Number of Hops) for **(a)** 25×25 Grid **(b)** 50×50 Grid.

T_v. It was shown in [5] that when $r = \sqrt{\frac{c \ln n}{\pi n}}$, for $c > 1$, the cover time of $G(\mathcal{X}_n; r)$ for the blind random walk is asymptotically $c \ln\left(\frac{c}{c-1}\right) n \ln n$.

For $\epsilon \in (0, 1)$ define also the *ϵ-approximate cover time* as the mean number of steps that $\mathcal{W}$ needs in order to visit a fraction $1 - \epsilon$ of the vertices. More formally, define $T_v^{(\epsilon)}$ to be the time needed until $\mathcal{W}_v$ has visited $(1 - \epsilon)n$ vertices. The *ϵ-approximate cover time* C of $G(\mathcal{X}_n; r)$ is defined as $C^{(\epsilon)} = \max_{v \in V} \mathbb{E}[T_v^{(\epsilon)}]$.

These cover times are related to latency, as they capture the time the sink needs to collect the sensory data from the entire network. The approximate cover time metric is of great interest as the majority of overlaps occur while the sink tries to locate the last few unvisited sub-regions that are scattered in the network area; however, in most sensor network applications it is sufficient to collect a vast percentage of the total sensory data, so this metric is relevant and informative.

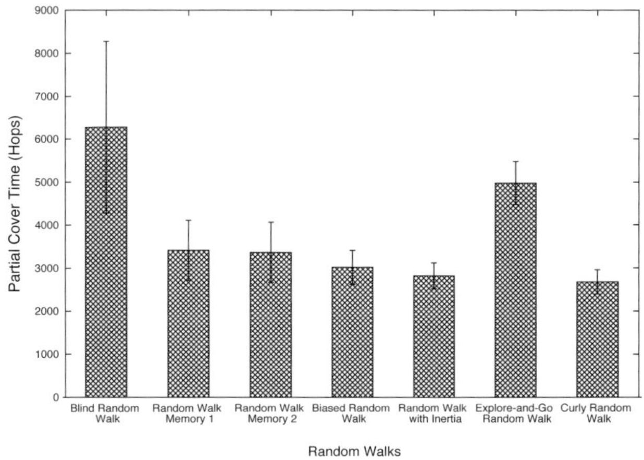

(a)

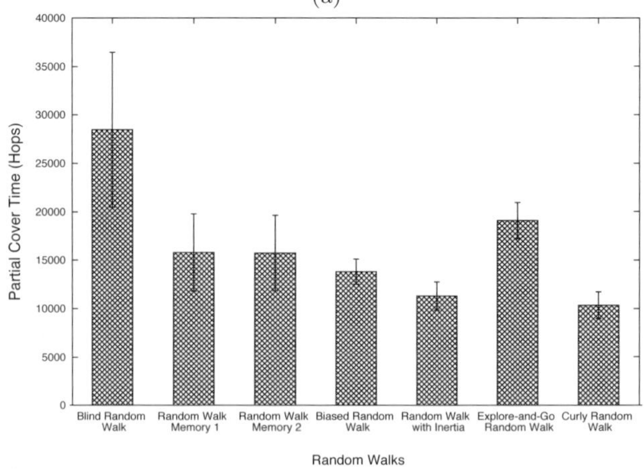

(b)

Fig. 2.5 Partial Cover Time (Number of Hops) for **(a)** 25×25 Grid **(b)** 50×50 Grid.

2.5.2 *Proximity Cover Time*

In real-life sensor networks, the sink is capable of collecting data not only from the currently visited node, but also from all nodes that are inside the communication range r (via a single-hop data transmission). Therefore, there is a strong motivation from real-life WSN to consider as visited not only the vertex of the current state of the walk, but also all neighboring vertices. Following is the formal definition of the corresponding metric, Proximity Cover Time.

Denote by $D(v, \rho)$ the disc of radius $\rho > 0$, centered at v. Consider the random walk $\mathcal{W}_v$ that begins on vertex v of a random instance of the random geometric graphs model $G(\mathcal{X}_n; r)$. We will say that $\mathcal{W}_v$ is *within distance* ρ to vertex u at time t if $W_v(t) \in D(u, \rho)$, i.e., the random walk started at v occupies a vertex that lies in $D(v, \rho)$ at time t. Let $T_v(\rho) \stackrel{def}{=} \inf\{t|\forall u \in V, \exists t' \leq t : W_v(t') \in D(u, \rho)\}$ be the time needed until $\mathcal{W}_v$ has come within distance ρ to all vertices of the graph. We define the *ρ-proximity cover time* of $G(\mathcal{X}_n; r)$ as $C(\rho) = \max_{v \in V} \mathbb{E}[T_v(\rho)]$.

A similar but different metric is presented in [3] where each time the random walk visits an already visited vertex, then u.a.r. picks an unvisited neighbor (if one exists), marks it as visited and then continues the random walk from its current vertex (i.e., it does not make a transition to the marked neighbor).

2.5.3 *Proximity Variation*

For a vertex $v \in V$ and time $t_0 \geq 0$ let

$$dist(v, t_0) = \min_{0 \leq t \leq t_0} \|\mathcal{W}(t) - v\|$$

be the function that returns the minimum distance between the random walk $\mathcal{W}$ up to time t_0 and node v. Then the *proximity variation* $PV(t_0)$ *at time* t_0, is given by

$$PV(t_0) = \frac{\sum_{v \in V} dist(v, t_0)}{n}$$

where n is the total number of nodes.

The rationale behind this metric is the following: In a real network, if the PV metric converges to zero quickly with t, this means that the sink gets close to all sensors quite soon and data collection progresses fast; this is especially relevant in case when the role of sensors is not completely passive but includes some limited multi-hop propagation of data to accelerate data propagation at a reasonable energy cost. On the contrary, when the PV converges slowly to zero, it means that the network traversal is performed in a way that some areas may stay unvisited for long time.

2.5.4 *Visit overlap statistics*

Other metrics that will be used in order to further characterize the evolution of each walk include the number of visits to a specific vertex v, as well as the distribution of the number of visits for every vertex in the graph. These overlap statistics will provide an insight on how exactly each walk traverses the graph, how different components of the graph are visited and in what rate.

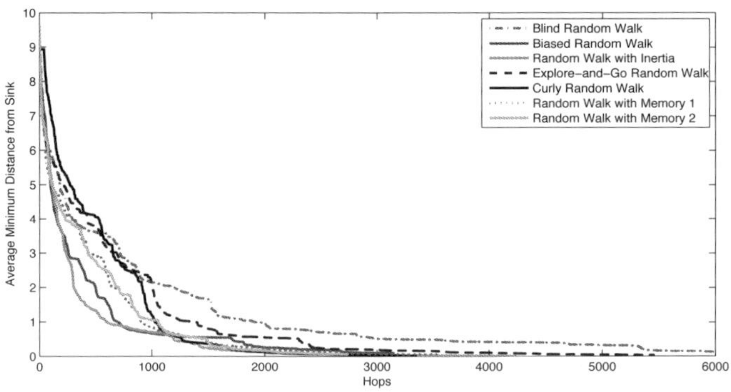

Fig. 2.6 The Proximity Variation for a 25 × 25 network.

2.6 Performance Evaluation

2.6.1 *Evaluation on the Grid model*

2.6.1.1 *Simulation set-up*

We used Matlab R2009a as our simulation environment. We evaluate the walks for two network dimensions, 25×25 and 50×50 square cells. We also select a random uniform distribution of the sensors over the network area. The width of each cell corresponds to one length unit. For the partial cover time metric, we set the percentage of the network to be covered to 95%.

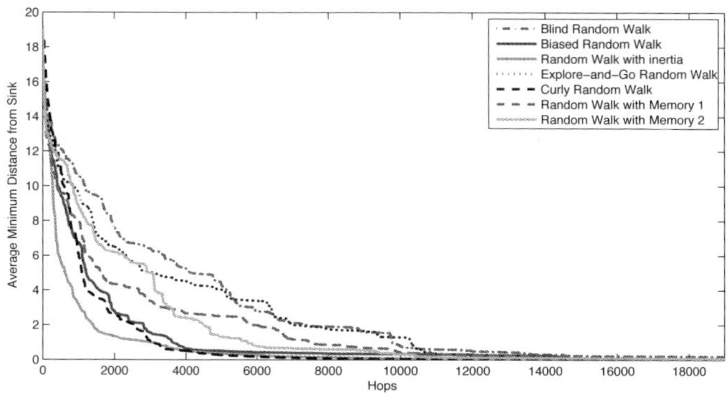

Fig. 2.7 The Proximity Variation for a 50 × 50 network.

For each network dimension we conduct 2500 iterations and compute the mean values for each metric, as well as the 90% confidence intervals of the findings. In the figures below, the top ends of the bars indicate observation means and the line segments represent the confidence intervals surrounding them. One can observe that our measurements proved statistically smooth in the sense that there a is high concentration around the mean values.

2.6.1.2 *Simulation findings*

The performance of the walks in terms of *cover time*, for two different network sizes, is depicted in Fig. 2.4. As shown in the figure, the Blind Random Walk has the worst performance among the seven mobility strategies. Using memory of size one or two improves the performance a lot, as we can see in the figure. The Random Walk with Inertia and the Explore-and-Go Random Walk perform significantly better than the Blind Random Walk and their behavior is comparable with the Random Walks with Memory One and Two. In fact, the Random Walk with Inertia is 45% better than the Blind Random Walk, while the Explore-and-Go Random Walk is 20% better than the Blind Random Walk for the 25×25 grid (Fig. 2.4a) and 40% better for the 50×50 grid (Fig. 2.4b). Furthermore, one can see that the performance of the Curly Random Walk is even better than the other two proposed mobility strategies. Especially in the case of the 50 × 50 network (Fig. 2.4b) it even outperforms the Biased Random Walk which is expected to have the best performance, since it optimizes its trajectory by remembering the whole history of visits in each cell. Furthermore, the performance of the Random Walk with Inertia is only about 20% worse than the performance of the Biased Random Walk, even though it only uses $O(1)$ memory.

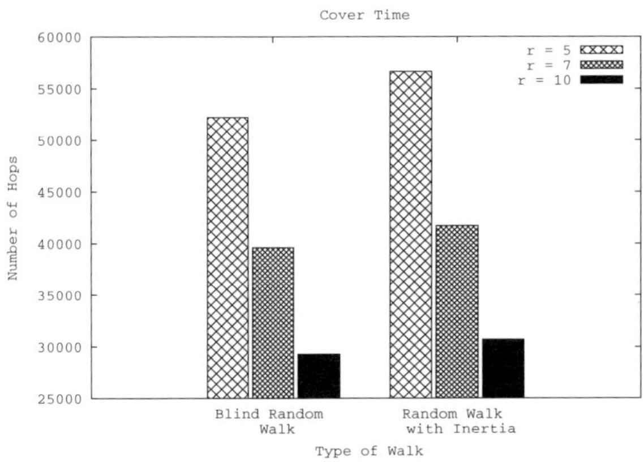

Fig. 2.8 Cover Time of the Blind Random Walk and the Inertia Random Walk.

Figure 2.5 shows the *partial cover time* of the seven walks, for the two grids. We observe that the Random Walk with Inertia and the Curly Random Walk cover the 95% of the network faster than the other walks (the Curly Random Walk is slightly better). These two walks perform better even than the stronger Biased Random Walk. The Explore-and-Go Random Walk is faster than the Blind Random Walk, too.

Figure 2.7 depicts the mean value of the smallest distance from the sink for all the cells over the total number of hops the sink has made for a network of 50×50 size. We compute the evolution of this metric until the entire network area is covered. Since the initial position of the sink for all the experiments is the center of the network area, at the beginning of each experiment this metric has the same value for all the walks. Also, because of the fact that each experiment is concluded when the whole network area is covered (all nodes are visited at least once), this metric converges to zero towards the end of each experiment.

However, because of the different way each walk traverses the network, significant differences are observed mainly in the speed with which zero is approximated. In particular, the Blind Random Walk converges very slowly to zero, followed by the Explore-and-Go Random Walk. This is because these two walks tend to move slowly inside the network area by frequently changing direction. On the contrary, the Random Walk with

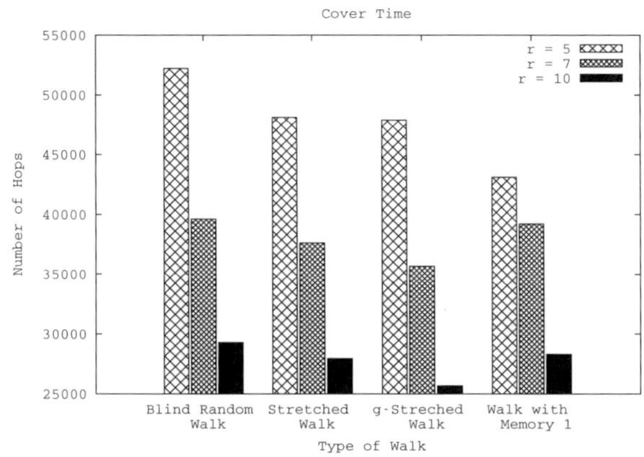

Fig. 2.9 Cover Time of the Blind Random Walk, Stretched Walk, γ-Stretched Walk and Walk with Memory 1.

Inertia converges to zero very quickly because of the long paths it creates during the network traversal. These long paths allow the sink to quickly approach, within a few hops distance, cells of different sub-regions. The Curly Random Walk converges to zero very fast, too, since the circular expansion of the movement of the sink on the network area results to a quick approach of different sub-regions.

2.6.2 *Evaluation on the Random Geometric Graphs model*

2.6.2.1 *Simulation set-up*

We conducted our experimental evaluation using Matlab R2009a as our simulation environment. We evaluate the walks on several instances of the Random Geometric Graphs model in order to address its random nature. More particularly, we construct Random Geometric Graphs consisting of $n = 2000$ nodes placed in a 100×100 square area. For such RGG graphs, the connectivity threshold is computed to be of radius $r = 100\sqrt{\frac{\ln 2000}{\pi 2000}} = 3.47$. For our performance evaluation we use three characteristic values for r: $r_1 = 5$, $r_2 = 7$, $r_3 = 10$, corresponding to sparse, average and denser graphs respectively. For every value of r, we construct 50 RGG instances on which we evaluate all walks for 100 iterations per instance. We consider these randomly constructed RGG instances to be representative of the RGG space.

For the performance evaluation of the protocols we used two network topologies. At the first topology sensor nodes are deployed uniformly at

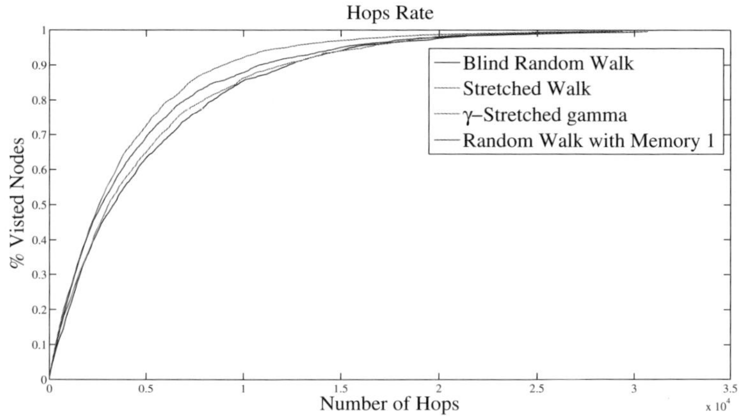

(a) Nodes' Discovery Rate for $r = 5$

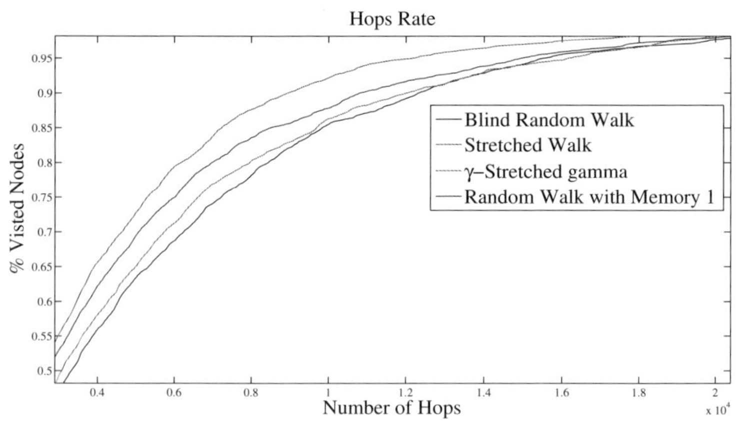

(b) Nodes' Discovery Rate for $r = 5$ (zoom-in)

Fig. 2.10 Percentage of discovered nodes over number of hops with range $r = 5$ and uniform distribution.

random over the planar area. This topology constitutes the base of our performance evaluation study by providing a first insight on the performance of each protocol. At the second network topology a portion of the sensors (in our experiments 60%) is deployed uniformly at random in order to establish connectivity. The rest of the sensors are deployed over the network area following the normal distribution with mean value $\mu = 50$ (i.e., half the dimension of the network area) and standard deviation $\sigma = 7$ for the x and y co-ordinates (independently of each other) of each sensor.

This set-up leads to a heterogeneous network deployment in which a high concentration of sensor nodes is located around the center of the network area.

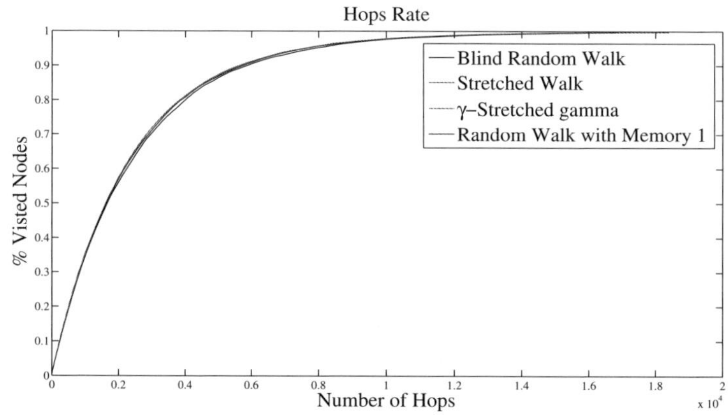

(a) Hops Rate for $r = 10$

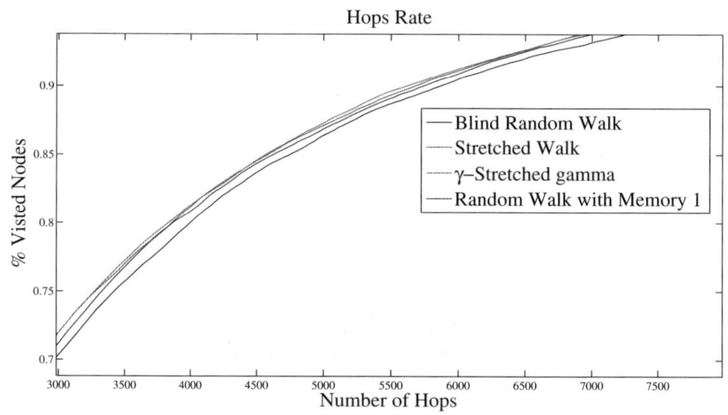

(b) Hops Rate for $r = 10$ (zoom-in)

Fig. 2.11 Percentage of discovered nodes over number of hops with range $r = 10$ and uniform distribution.

2.6.2.2 *Simulation findings*

We first experimentally evaluate the performance of the Random Walk with Inertia on the Random Geometric Graph and compare it with the Blind

Random Walk (only in the uniform topology). Then we compare four random walks; the well known Blind Random Walk; two versions of our newly proposed walk: the Stretched Walk where $\gamma = 1$ and the γ-Stretched Walk, where $\gamma = \frac{r}{r_c}$, in order to investigate the impact of γ; the Random Walk with Memory 1, that somehow represents an upper bound in terms of memory usage and therefore is considered to be more powerful than the rest of the walks.

2.6.2.3 *Uniform topology*

Figure 2.8 presents the comparison between the Random Walk with Inertia and the Blind Random Walk. Although the Random Walk with Inertia on the Grid performs better than the Blind, however the RGG model causes the Inertia to present a really weak performance. This result led us to the design of the newly proposed random walk, the so called γ-Stretched Random Walk.

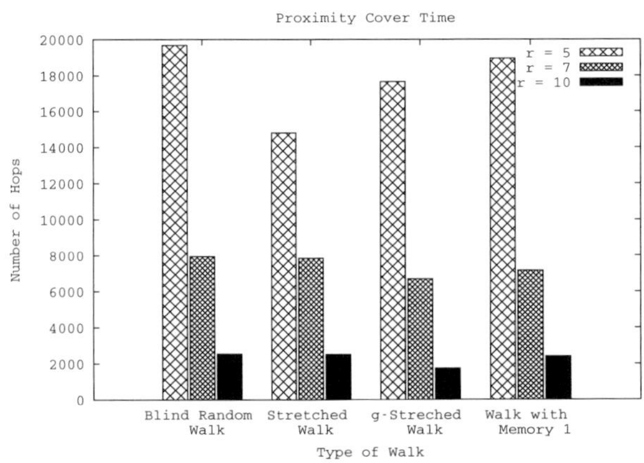

Fig. 2.12 Proximity Cover Time.

Figure 2.9 depicts the mean number of hops each walk needs in order to visit all the nodes of the network for three representative values of the communication range r. For all the random walks we observe that as the density of the graph increases, the total cover time is reduced even by nearly 50%. This is due to the fact that in a dense network the walk has more edges to traverse on, and therefore there is higher probability to visit any given vertex. In other words, a random walk reaches its stationary distribution much faster. We also observe that for all graph densities, the Blind Random

Walk has the biggest cover time, while the impact of the fixed-sized memory of the Walk with Memory declines as the density of the graph increases. In fact, in sparse graphs, where the available options are limited, avoiding 1-hop overlaps by not visiting the last position significantly reduces the cover time. In our Stretched Walks, on the contrary, the impact of the γ factor is proportional to the number of neighboring nodes as the graph density increases. This is why for small values of radius r the performance of the Stretched Walks is very similar. However, in more dense graphs, biasing the walk towards the more distant neighbors seems to have a well noticeable effect on the speed the random walk traverses the graph.

Figures 2.10 and 2.11 depict the rate at which each walk discovers unvisited nodes inside the network area for range $r = 5$ and $r = 10$ respectively. For sparse networks ($r = 5$) we note that the Stretched Walk outperforms even Random Walk with Memory. The Random Walk with Memory, while avoiding its previous position, may move towards a neighboring node that geometrically is very approximate; thus allowing a high probability for early overlaps. On the other hand, the Stretched Walk, by favoring the more distant neighbors, quickly changes the subregions of the networks it traverses. This way, it avoids early overlaps, thus discovering unvisited nodes much faster. However, in sparse networks, if the bias factor towards distant neighbors is very crude it may have an opposite effect. As seen in the performance of the γ-Stretched Walk, a crude bias factor may narrow the available options for the next step down to the point the walk is forced to make long cycles during its traversal, thus visiting already visited nodes.

In dense networks ($r = 10$) the rate at which unvisited nodes are discovered is accelerated for all walks. However, γ-Stretched Walk performs slightly better as the bias factor strongly favors distant nodes over geometrically approximate ones. We note that for all walks while they all discover relatively quick the 95% of the total number of vertices in the graph, they spend nearly half of the cover time in order to visit the last 5% of the vertices, thus creating a long convergence tail at the end of the traversal.

As discussed in subsection 2.5.2, there is a strong motivation from real-life wireless sensor networks to consider as visited not only the vertex of the current state of the walk, but also all neighboring vertices in the RGG network. Figure 2.12 depicts the performance evaluation of all walks in RGG for the Proximity Cover Time metric. Intuitively, the dominant factor for this metric is the dislocation attribute of the motion (i.e., the mean geometric distance the sink covers per hop). In sparse networks the use of memory by the Random Walk with Memory slightly improves the Proximity

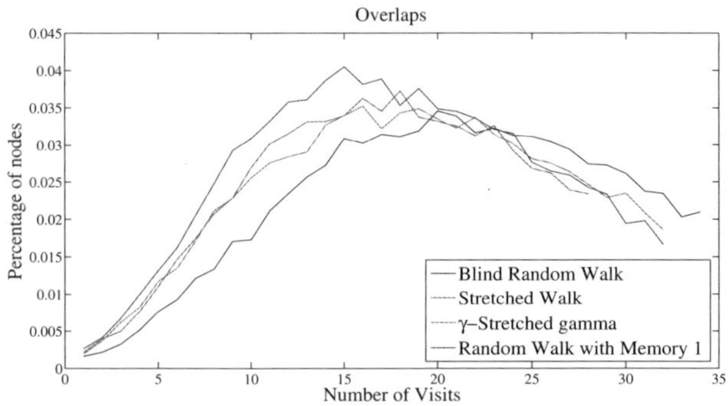

(a) Overlaps distribution for $r = 5$

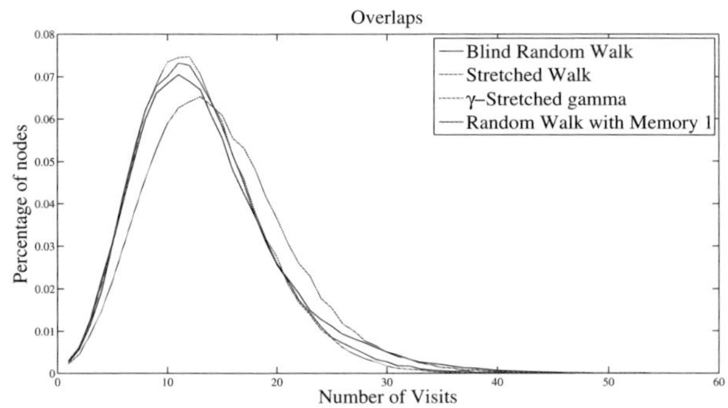

(b) Overlaps distribution for $r = 7$

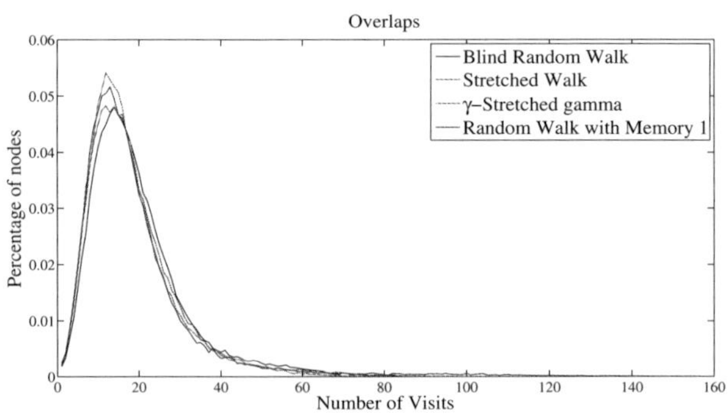

(c) Overlaps distribution for $r = 10$

Fig. 2.13 Overlaps distribution over the nodes of the graph for the random uniform topology.

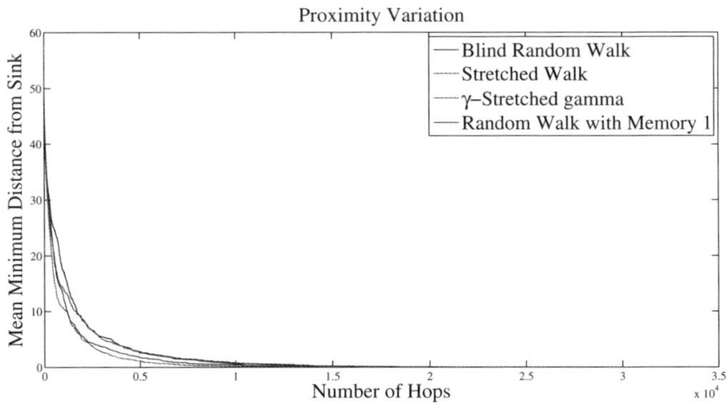

(a) Proximity Variation for $r = 5$

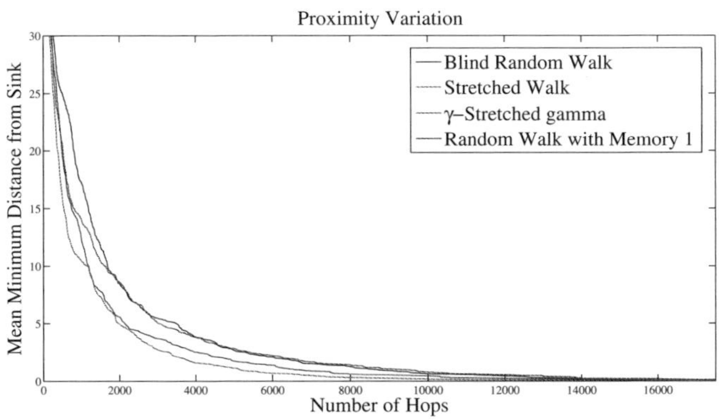

(b) Proximity Variation for $r = 5$ (zoom-in)

Fig. 2.14 The Proximity Variation Metric for $r = 5$ and uniform topology.

Cover Time of the Blind Random Walk. The Stretched Walk with $\gamma = 1$ shows the best performance from all other walks. The γ-Stretched Walk is outperformed due to the crude way it is biased towards the few distant neighbors. However, for higher network densities this factor aids the walk to outperform even the Walk with Memory. In a network consisting of 2000 nodes, in less than 2000 hops the sink has been inside the communication radius of every node in the network.

Figure 2.13 depicts the overlaps distribution over the nodes of the graph. That is what percentage of the total number of the nodes has been visited i times by the sink during the entire network traversal process. In sparse graphs the Random Walk with Memory outperforms the two Stretched

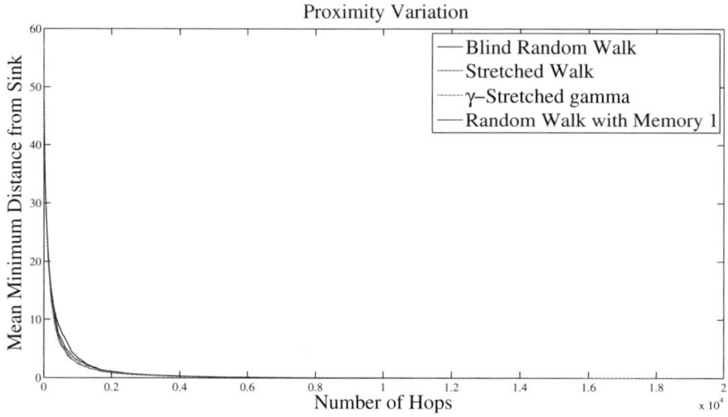

(a) Proximity Variation for $r = 10$

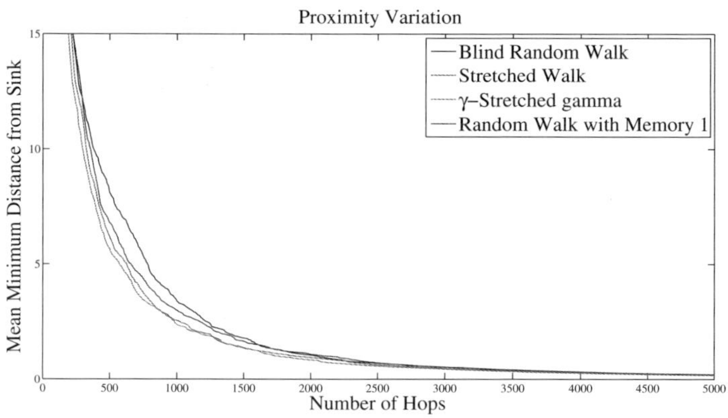

(b) Proximity Variation for $r = 10$ (zoom-in)

Fig. 2.15 The Proximity Variation Metric for $r = 10$ and uniform topology.

Walks, that is more nodes of the network have been visited fewer times (i.e., the curve corresponding to Walk with Memory is above and to the left than the rest). At first glance this contradicts the fact that the Stretched Walk is discovering unvisited nodes in sparse networks at a higher rate than the Walk with Memory (Fig. 2). We note however that the Stretched Walk avoids early overlaps, while it needs more time to cover the entire network (Fig. 2.9). This leads to the conclusion that most of the overlaps occur towards the end of the network traversal process, thus the converging tails in Fig. 2.

For denser networks the impact of the bias factor of γ-Stretched Walk increases along with the value of r, thus eventually demonstrating the best performance for this metric. Also, as the network graph density increases, the number of revisits for the majority of the nodes is more or less the same ($10 - 15$ revisits). However, the total number of revisits is increased significantly, as the last few unvisited nodes are hard to find in a dense network.

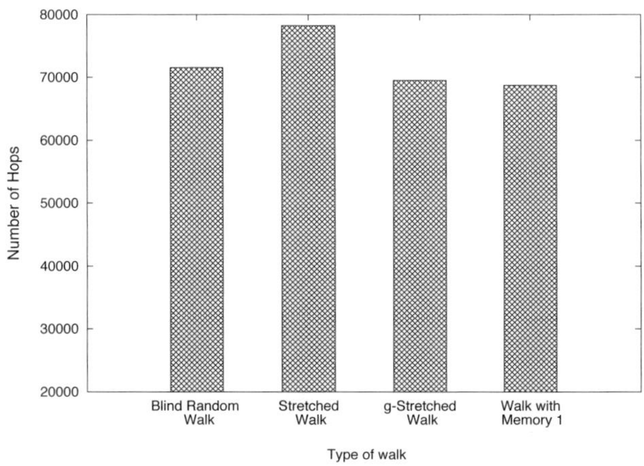

Fig. 2.16 Cover Time for heterogeneous topology.

There lies a different intuition behind each walk on how the network area is traversed, regardless of the total number of hops needed. For example the strategy of avoiding previous positions is getting less effective when the density of the network increases, while the γ factor in Stretched walks is allowing the sink to visit many different network subregions very early in the traversal process. The proximity variance metric captures these differences. In Fig. 6 we note that the two walks which perform better in sparse graphs are the Stretched Walk (where $\gamma = 1$) and the Random Walk with Memory. As there are relatively few neighboring nodes slightly favoring distant nodes or just avoiding previous position seems to suffice. On the other hand, the γ-Stretched Walk (where $\gamma = \frac{r}{r_c}$) favors distant nodes in a cruder manner (for $r = 5$ and $r_c \approx 2.3$, $\gamma \approx 2$). This way, the walk is left with very few available options, forced to make several cycles during its traversal process.

In dense networks, where there are many available options for the next step of the walk, this crude selection of the distant nodes allows the walk not to get attracted by approximate neighbors. Therefore, in Fig. 2.15 the

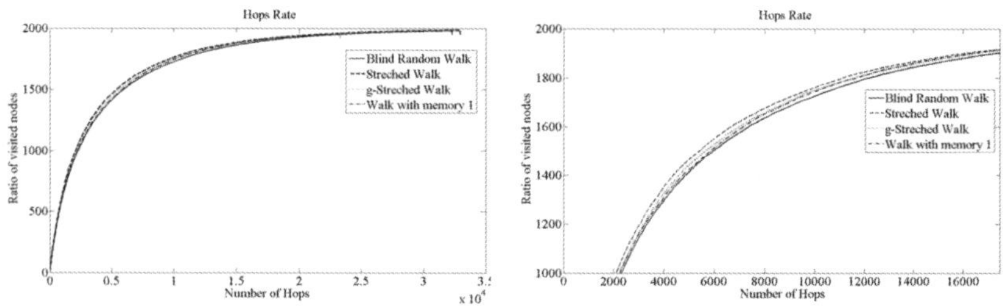

(a) Hops rate for $r = 7$ (b) Hops rate for $r = 7$ (zoom-in)

Fig. 2.17 The Hops rate for $r = 7$ and heterogeneous topology.

γ-Stretched Walk is the one converging to zero at a higher rate.

2.6.2.4 *Normal distribution*

Figure 2.17 depicts the cover time of each walk for a specific characteristic communication range r. As shown in the figure, the γ-Stretched Random Walk performs better than the Blind Random Walk. The γ biased factor allows the Stretched Walk not to get trapped inside dense areas of the network, contrary to the Blind Walk. On the other hand, the simple Stretched Random Walk has the worst performance.

A better insight on the performance of our two proposed walks for the normal distribution can be obtained by analysing their performance over the uniform network topology shown in Fig. 2.9. We notice that in sparse areas both walks have similar performance regarding their cover time and the impact of the γ factor is negligible. This is due to the fact that in sparse areas both walks have few alternatives for their next move. On the contrary, in dense areas the γ factor allows the γ-Stretched Walk to quickly traverse the network area by favoring the more distant neighbors resulting in better performance. As a result, in the case of a heterogeneous topology, the γ-Stretched Walk behaves similarly in the sparse areas of the network, but much better in denser areas, yielding a better overall performance.

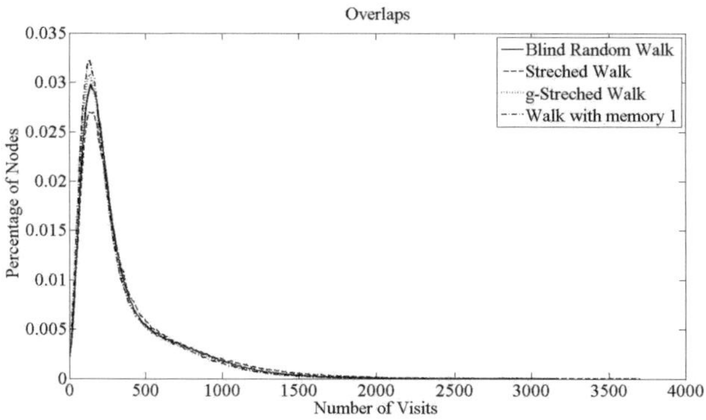

Fig. 2.18 The overlaps distribution for $r = 7$ and heterogeneous topology.

Figure 2.17 presents the rate at which each walk discovers unvisited nodes inside the network area. We note that the γ-Stretched Random Walk outperforms even the Random Walk with Memory. As in the uniform topology, the Stretched Walk, by favoring the more distant neighbors, quickly changes the subregions of the networks it traverses, this way avoiding early overlaps and discovering unvisited nodes much faster.

Figure 2.18 depicts the overlaps distribution over the nodes of the graph. That is what percentage of the total number of the nodes has been visited i times by the sink during the entire network traversal process. One can see that the Random Walk with Memory outperforms the two Stretched Walks, that is more nodes of the network have been visited fewer times. This does not contradict the fact that the γ-Stretched Walk discovers unvisited nodes at a higher rate than the Walk with Memory. We note however that the Stretched Walk avoids early overlaps, while it needs more time to cover the entire network. Actually, most of the overlaps occur towards the end of the network traversal process, especially in this heterogeneous topology, since it is more difficult for the Stretched Walks to discover the nodes of the dense area thus leading to multiple overlaps.

Finally, Fig. 2.19 depicts the proximity variation metric for each walk. As shown in the figure, the Stretched Random Walk ($\gamma = 1$) has the best performance, since it favors distant neighbors, but is not that biased as the γ-Stretched, thus succeeding to visit many different network subregions, both in the sparse but also in the dense areas of the network. On the other hand, the γ-Stretched Walk (where $\gamma = \frac{r}{r_c}$) favors distant nodes in a more crude manner (for $r = 7$ and $r_c \approx 3.5$, $\gamma \approx 2$). This way, the walk is left with very few available options, forced to make several cycles during its traversal process.

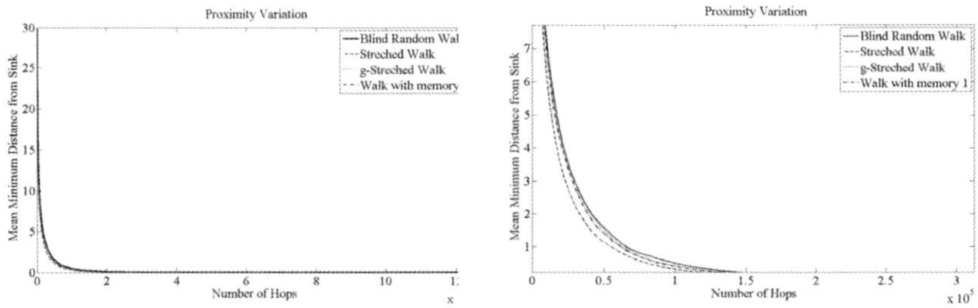

(a) Proximity Variation for $r = 7$

(b) Proximity Variation for $r = 7$ (zoom-in)

Fig. 2.19 The Proximity Variation for $r = 7$ and heterogeneous topology.

2.7 Conclusions

In this chapter, we address the problem of efficient data collection in wireless sensor networks via a mobile sink. We model the data collection process by random walks on two network models. At first, we study random walks on a Grid that is virtually overlaid over the network area. Its regular structure helps us design three new random walks; the Random Walk with Inertia, the Explore-n-Go Random Walk and the Curly Random Walk. Then, we adopt the Random Geometric Graphs model that better captures the geometric dependencies that characterize inter-node wireless communication in wireless sensor networks. On this model, we define a new walk called Stretched Walk and evaluate two variations of it. We evaluate all walks against well known random walks (blind random walk and random walks with memory) in terms of cover time, approximate cover time, overlap statistics, proximity cover time and proximity variation. The two latter, recently introduced metrics capture differences on the way each walk traverses the network area.

Findings show that the use of constant sized memory in order to accelerate the data collection process has limited results, particularly when the network density increases. Instead, walks that use local (immediate neighbor) information in a smart manner can demonstrate significantly better results. However, their use should be wise as they may lead to opposite results (e.g., strong bias factor in sparse networks leads to bigger cover times).

References

[1] Angelopoulos, C. M., Nikoletseas, S. E., Patroumpa, D. and Raptopoulos, C. (2011). A new random walk for efficient data collection in sensor networks, in *MOBIWAC*, pp. 53–60.

[2] Angelopoulos, C. M., Nikoletseas, S. E., Patroumpa, D. and Rolim, J. D. P. (2010). Coverage-adaptive random walks for fast sensory data collection, in *ADHOC-NOW*, pp. 81–94.

[3] Berenbrink, P., Cooper, C., Elsasser, R., Radzik, T. and Sauerwald, T. (2010). Speeding up random walks with neighborhood exploration, in *Twenty-First Annual ACM-SIAM Symposium on Discrete Algorithms (SODA)*.

[4] Chatzigiannakis, I., Nikoletseas, S. E. and Spirakis, P. G. (2001). An efficient communication strategy for ad-hoc mobile networks, in *DISC*, pp. 285–299.

[5] Cooper, C. and Frieze, A. (2008). Component structure of the vacant set induced by a random walk on a random graph, *Random structures and Algorithms* **32**, pp. 401–439.

[6] Dhillon, S. S. and Mieghem, P. V. (2007). Comparison of random walk strategies for ad hoc networks, in *The Sixth Annual Mediterranean Ad Hoc Networking WorkShop*.

[7] Dimitriou, T., Nikoletseas, S. E. and Spirakis, P. G. (2004). Analysis of the information propagation time among mobile hosts, in *ADHOC-NOW*, pp. 122–134.

[8] Dimitriou, T., Nikoletseas, S. E. and Spirakis, P. G. (2006). The infection time of graphs, *Discrete Applied Mathematics* **154(18)**, pp. 2577–2589.

[9] Gupta, P. and Kumar, P. (1998). *Critical power for asymptotic connectivity in wireless networks* (Stochastic Analysis, Control, Optimization and Applications, Boston).

[10] Jarry, A., Leone, P., Nikoletseas, S. and Rolim, J. D. P. (2010). Optimal data gathering paths and energy balance mechanisms in wireless networks, in *DCOSS*, pp. 288–305, best Paper Award.

[11] Kinalis, A., Nikoletseas, S., Patroumpa, D. and Rolim, J. (2009). Biased sink mobility and adaptive stop times for low latency data collection in sensor networks, in *GLOBECOM '09: Proceedings of the 52nd IEEE International Global Communications Conference*.

[12] Krishnamachari, B. (2005). *Networking Wireless Sensors* (Cambridge University Press).

[13] Leung, K. K. (2012). Modeling and optimization of vehicular wireless ad-hoc networks. in *The 15th ACM International Conference on Modeling, Analysis and Simulation of Wireless and Mobile Systems (MSWiM)*.

[14] Li, X., Nayak, A. and Stojmenovic, I. (2010). Sink mobility in wireless sensor networks, in *Wireless Sensor and Actuator Networks: Algorithms and Protocols for Scalable Coordination and Data Communication*, p. Wiley.

[15] Lima, L. and Barros, J. (2007). Random walks on sensor networks, in *WiOpt '07: 5th International Symposium on Modeling and Optimization in Mobile, Ad Hoc, and Wireless Networks*.

[16] Luo, J. and Hubaux, J.-P. (2005). Joint Mobility and Routing for Lifetime

Elongation in Wireless Sensor Networks, in *24th IEEE INFOCOM* (Miami, USA).

[17] Luo, J., Panchard, J., Piorkowski, M., Grossglauser, M. and Hubaux, J.-P. (2006). Mobiroute: Routing towards a mobile sink for improving lifetime in sensor networks, in *2nd IEEE/ACM International Conference on Distributed Computing in Sensor Networks (DCOSS)*, Vol. 4026, ISBN 3-540-35227-9, pp. 480–497.

[18] Nikoletseas, S. E., Raptopoulos, C. and Spirakis, P. G. (2006). The survival of the weakest in networks, in *WAOA*, pp. 316–329.

[19] Penrose, M. (2003). *Random Geometric Graphs* (Oxford University Press).

[20] Yoon, J., Liu, M. and Noble., B. (2003). Random waypoint considered harmful. in *Proceedings of Infocom '03*, pp. 1312–1321.

Chapter 3

Robot-Robot Coordination

Ivan Mezei, Milan Lukić and Veljko Malbaša

University of Novi Sad, Faculty of Technical Sciences, Novi Sad, Serbia

Abstract. *Problem of coordination among robots is addressed in the context of Wireless sensor and robot networks (WSRNs) in this chapter. The two traditional fields of wireless sensor networks (WSNs) and multi-robot systems meet in the WSRNs. This chapter considers coordination among robots aimed for application of networked robots to manage and improve functionality of sensor networks, in various tasks in which response to events detected by sensors is needed. Coordination problem consists of two parts: task assignment (or allocation) and task fulfillment. Most of the papers in the traditional field of multi-robots deal with generic problem statements with heavy assumptions (e.g., complete communication graph is assumed) when it comes to task assignment problem. Literature still lacks solutions for the task assignment scenarios with non-ideal or limited communication, especially in cases with several concurrent events. This chapter focuses on the task assignment problem as a part of coordination among robots in WSRNs starting from the simplest single-event scenarios to complex multiple-event scenarios and reviews possible solutions and current research challenges.*

3.1 Introduction

Research in the field of the two traditional fields of Wireless sensor networks (WSN) and Multi-robot systems (MRS) has been done extensively during the last two decades [9, 24]. Wireless sensor and robot networks (WSRNs),

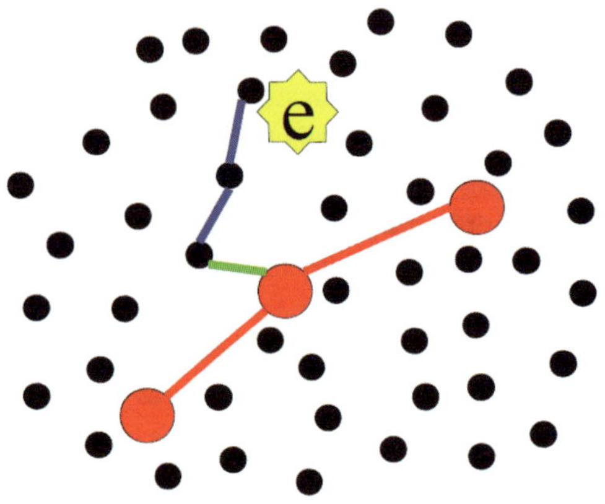

Fig. 3.1 Communication paradigms in WSRNs.

as an extension of Wireless sensor and actuator networks [2], is the confluence point where WSN and MRS research fields meet. It is a rather young field of research and there are still opened issues to be resolved [5, 7].

In the WSRNs, there are three types of communication paradigms that are illustrated in Fig. 3.1. Sensor-sensor communication (blue line) [2] is used to report event along the path towards a robot. Sensor-robot (green line) is used to report to a particular robot, while robot-robot communication (red line) is used for communication among robots. According to [7] communication among entities in MRS is a fundamental aspect of coordination.

There are four types of architectures in WSRNs: not-automated, semi-automated, cooperative and automated [2, 22]. Non-automated architecture is completely centralized and all decisions are made by the user of the network. In semi-automated networks, there is always one leading node or controlling unit that makes all decisions. Cooperative networks have the same architecture as semi-automated but usually include other nodes in the decision making process. In this chapter, we consider only automated networks where all decisions are made locally by robots in the event neighborhood.

There is a lot of research work done regarding the robot coordination problem (recent examples are [23, 27]). However, literature still lacks solutions for a number of problems in this research area, especially in the context of WSRN with communication costs included and with multiple concurrent event scenarios.

The robot coordination problem consists of two parts: task assignment (also known as task allocation [21]) and task fulfillment. Here we focus on the task assignment problem in different scenarios and review possible solutions and current research challenges. Scenarios, types of events and examples of tasks that can be assigned to robots can be different starting from the simplest single-event scenarios to the complex multi-event multi-robot scenarios (see Fig. 3.2).

Events	single multiple non-concurrent multiple "group by group" multiple concurrent
Tasks	single robot multiple robot
Task assignments	single task assignment to robot(s) multiple tasks assignment to robot(s)
Type of solution	centralized localized distributed market-based

Fig. 3.2 Different types of events, tasks, assignments and solutions.

Robots are always capable of performing only one task at a time but in some scenarios, they can be assigned with more than one consecutive task, since sometimes, tasks need more than one robot to be successfully accomplished (e.g., exhaustion of larger fire, heavy box pushing, coordinated exploration etc.). In the case of multiple events, it is important how events occur and if there is time bound for the task completion. If there is no time bound for task completion (or it is not an issue), all multiple event problems can be treated as iterated version of single event problem. Hence, multiple event problems differ from single event problems only in emergent applications.

Example of the simplest multiple task assignment problem is fire exhaustion scenario in which more than one location need to be exhausted by the same robot. However, it is more likely that, if there is more than one robot, every fire location is assigned to different robot or robots. A scenario with three concurrent fire locations and different assignment types is depicted in Fig. 3.3. Simultaneous deployment of new sensors to different

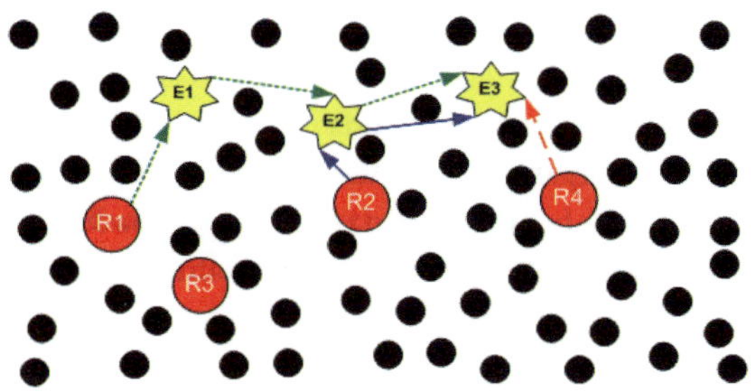

Fig. 3.3 Three concurrent events (E1, E2, E3 fire locations) and three possible assignments of robots (R1, R2, R3, R4 initial positions of robots).

parts of the area monitored by robots would be another multiple assignment application example.

For a given set of sensors, robots, events, available resources, costs and environment characteristics, the task assignment problem (the problem) within robot-robot coordination can be defined as follows: find the best robot(s) to respond to occurred event(s) and assign task(s). The problem objective would be to minimize costs within energy budget, since sensors and robots are resource limited (e.g., usually battery powered). This problem consists of two phases. First is to find the robot(s) starting from the sensed event. It is a non-trivial task because of robot mobility which causes constant topology changes. In order to find a robot from a sensor location, in case when exact locations of robots are unknown, a discovery service (also known as location service) must be used [14]. Discovery service is in fact a difficult problem if the communications are not ideal. The second task is to find the best robot(s) to (re)act upon sensed event(s).

There are centralized, distributed, localized or market-based solutions for the task assignment problem. Several constraints could have a significant influence on the task assignment. Robot energy, time needed to start an action (responsiveness) or time needed to complete an action could be important constraints. Centralized solutions assume that one of the robots is chosen to be the leader and it gets all necessary data from sensors, makes decision and performs task assignment. In this way, theoretically, the leader can make optimal plan since it is supposed that the leader has all relevant information for decision making. However, it is computationally hard, has low fault tolerance with heavy communication load, possibly incurring a lot of collisions, especially with large number of robots. It also implies bad scalability and slow responsiveness for large networks. Distributed and localized solution utilizes spreading all planning responsibilities among robots that are in the event neighborhood. Robots use only locally available information to make their plan. Basic characteristics are: good responsiveness, low computation and communication costs, robustness, good scalability and fault tolerance. However, assignments could be highly suboptimal due to localized decisions. Market based solution is a kind of distributed solution (sometimes called semi-distributed). It is based on an economic model of market. Robots are working as free agents, negotiate and bid for tasks by means of auctions, to maximize their individual profit. The goal is global efficiency by completing all tasks while minimizing all costs. The main contribution of market based approach is that it tends to achieve positive effects from centralized solutions while minimizing negative ones.

There are several issues regarding an optimal solution for the task assignment problem. If the algorithm for discovery service is good, it should find the closest robot. It seems that the closest robot is the best candidate to be assigned. But is it so? What if that robot hasn't enough remaining energy to come and exhaust the fire? By means of robot-robot communication, it should be decided which one should act. What if there is more than one event? On Fig. 3.3, three possible assignments are depicted (one robot visits all, one robot per event and mixed). What methods can be used in such scenarios?

Starting from the single-event scenarios (e.g., fire exhaustion) followed by the multiple-event scenarios (version with three events is depicted on Fig. 3.3), several different scenarios will be considered. Solutions to the task assignment problem will be explored and the taxonomy of task assignment methods for those scenarios with special accent on communication which are considered ideal or non-ideal.

3.2 Single-event Scenarios

Single-event scenarios are the simplest task assignment problem to be solved within robot-robot coordination, especially if communication costs are not included. It is assumed that every robot has information about its geographic position (e.g., robot has GPS device integrated) and geographic position of the event is also known to the nearest sensor (by means of detecting sensor localization or sensors could have low power GPS integrated). Robots are capable of performing only a single task at a time. It is assumed that there is always at least one sensor inside robot communication range (making sensor-robot communication always possible).

First, we will explore scenario in which communications are considered ideal thus incurring no cost. In the majority of papers, especially in the field of multi-robot systems, the complete graph of the network is considered. It is a strong assumption that is feasible only for small networks. However, even in the small networks, it would incur high rate of collisions and retransmissions. Additionally, a priori knowledge of the event is known (e.g., time of event occurrence, location of event, duration etc.). We assume here only that the communication incurs no cost, and coming event is not known in advance. Accordingly, all costs are due to mobility of the robots. Upon sensed event, sensors route information about event to the nearby robot. In scenarios where sensors already know the location of all nearby robots, a sensor could decide the closest robot or the best robot to act upon event. In scenario when nearby locations are not known in advance, first step is to flood the whole network to obtain all robot locations. First response could be taken as the best (good responsiveness) or that robot could consult others to see if there is a better one in the vicinity (negotiation based coordination). Another solution is when all robots start to move towards event area, but stop when first one arrives at the location and flood others that it already was assigned with task.

If robot energy resources are limited, making decision which is the best robot is made by assigning task to the nearest robot that has enough remaining energy to come to event area and accomplish the task. In this case, this may not necessary be the nearest robot to the event. Making decision in this scenario can be made by means of negotiation between robots. The purpose of negotiation could be to find the nearest robot (low energy consumption for team). If the goal is to maximize the time the first robot looses all its energy then a different strategy that balancing load among robots is needed. In robotic community, there is a lot of research work done in the field of robot coordination using market-based approach [6]. A possible application of market-based approach in this scenario is as follows. Basic idea is that each robot calculates the distance to the event or some other value (e.g., $m \times \frac{d}{R_E}$ where m is a constant, d distance from the event, and R_E is remaining energy), and accordingly responds by bidding on auction organized by the sensor or by the robot that first receives event information. The robot that sends the best bid will be assigned the job. Here we can determine a localized solution where only nearby robots participate in the auction, while centralized auction is for all robots. In all previous scenarios, communication costs are neglected. In real situations, communication costs are significant and should be taken into account. Accordingly, a discovery service incurs significant cost and is an important issue within this class of scenarios. Upon event occurrence, a sensor could route the event information towards robots, but the question is how to do it if robot locations are not known in advance. The simplest idea is to flood the whole network to route information. Flooding has communication complexity of $O(n)$, where n is the number of nodes. The best flooding algorithm (if such exists), where each message is retransmitted only once and each node receives only one copy of message, total cost in terms of hop-count would be the number of nodes increased by the number of robots. Hence, flooding incurs high communication costs and limits bandwidth usage. Localized or limited flooding with hop-count restriction may be used as mean to lower the communication cost, but delivery can't be guaranteed (e.g., all robots are distanced from event area more than hop-count limit). It might still incur high communication overhead. Since flooding features high communication costs, mainly due to not knowing locations of the nearby robots, other solutions must be considered. Algorithms and protocols that are finding the nearby or the closest robots belong to the class of service discovery problems. Quorum-based and other service discovery methods are given in [22]. Quorum-based service discovery features good scalability, high location retrieval success rate, almost guaranteed delivery and complexity of $O(\sqrt(n))$.

Quorum-based service discovery significantly decreases communication overhead compared to flooding. However, the probability of really finding the closest service provider (robot) can still be improved by constructing information mesh (iMesh), as explained in [14]. In the iMesh structure construction phase, all the robots send the information about their positions through the static sensor network in four directions (north, east, south and west), by sending registration messages. When two registration paths collide, a blocking rule is applied, by which a registration message which is further from its source stops its propagation, while the other one goes on. The nodes along the path of registration messages keep the information about positions of robots that initiated them, thus forming the iMesh structure (solid lines in Fig. 3.4). The iMesh structure is used for service discovery. When an event occurs, the sensor that senses it sends cross-lookup requests in four geographic directions, and they propagate in a similar fashion as registration messages. When cross-lookup reaches some of the nodes within the iMesh structure, that node generates cross-lookup answer containing the information about service provider that is sent back to the node that initiated the request. When the event-sensing node collects

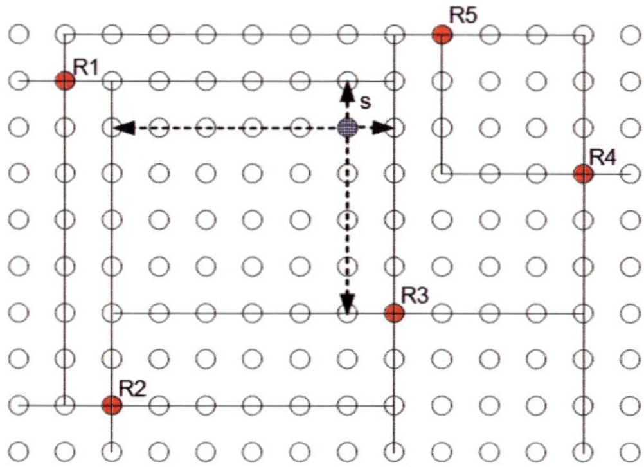

Fig. 3.4 iMesh cross-lookup doesnt find the closest robot.

all the cross-lookup answers, it determines the location of the nearest service provider with high degree of accuracy. However, there are scenarios when iMesh structure fails to give information about the closest robot as depicted in Fig. 3.4. Here, the closest robot to the sensor s is R5, but iMesh cross-lookup will give R3 as the closest.

Solutions that reduce communication overhead within robot-robot coordination are localized auction aggregation protocols (k-SAP, k-SAAP, k-AAP) [19]. The collecting robot may itself have a low cost of performing needed action. It includes its own cost in message that is flooded to other robots. Robots receiving the service request will compare their own cost and attach to the message they retransmit only the lowest cost (between incoming one and their own). The specific protocol has tree expansion and tree contraction phases.

Tree expansion starts from collecting robot R1, creating a tree rooted at R1 (see Fig. 3.5(b)). Each remaining robot, when the request was received for the first time, retransmits it with the lower cost appended to the packet. These retransmissions create a response tree. Each node with retransmission includes its parent robot in the message, so that robots can decide whether or not they are leaves in the created tree. They become leaves if they do not hear any other robot listing them as their parent. Leaf nodes start responding back to robots, with best cost they are aware of. Each intermediate node waits to hear from all neighbors that declared it as parent. In the scenario in Fig. 3.5, R2 and R3 wait for response from all children robots. After hearing, they select the best cost and report further towards the collector. The collector then decides who the best to perform required action is, and routes the task to that node. In fact each subsequent parent node organizes a local auction for its children. This methodology generates aggregation effect and is an improvement over the simple auction protocol [17], where each node responds to the parent node using separate routing paths, as depicted on Fig. 3.5(a). Overall effect is that total number of communication messages is reduced at a cost of small decrease in decision optimality. It is shown that for k-SAAP protocol, the number of sent messages per robot is always less or equals two (upper bound) [18].

To improve suboptimality of localized decisions of localized auction aggregation protocol (k-SAAP), especially for large networks and low values of k, 1-hop greedy search for better robot could be added after the original assignment is proposed. That is, the selected robot, after getting the task, will check if any of its neighbors has a lower cost. If so, it can reassign the task, and such a search can be repeated by the selected neighbor, until no improvement is possible.

To enhance k-SAAP, we proposed the following greedy improvement. After k-SAAP protocol is finished and (possibly) the best robot is selected, 1-hop SAAP can be initiated by the previously decided robot in order to search for a better robot in the 1-hop neighborhood. If there is no such robot, no reassignment is performed. If a better robot exists, it becomes new collecting robot and initiates another 1-hop SAAP. The algorithm (designated as k-SAAPG) repeats until no better robot is found in 1-hop neighborhood. This algorithm is published in the context of wireless sensor and actuator networks in [20].

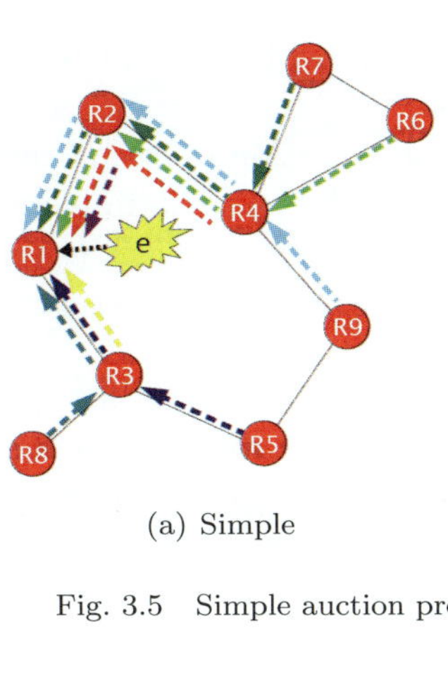

(a) Simple

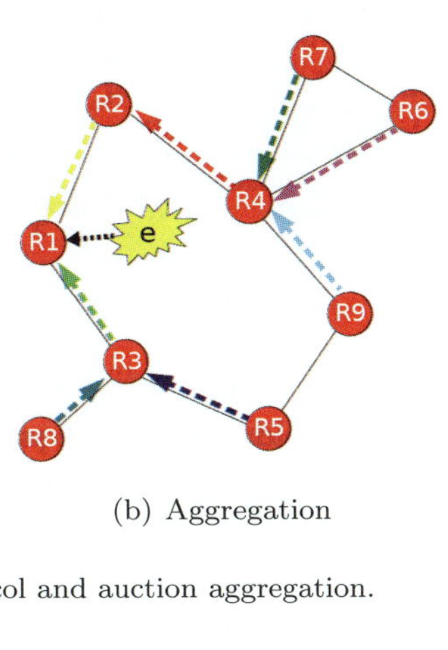

(b) Aggregation

Fig. 3.5 Simple auction protocol and auction aggregation.

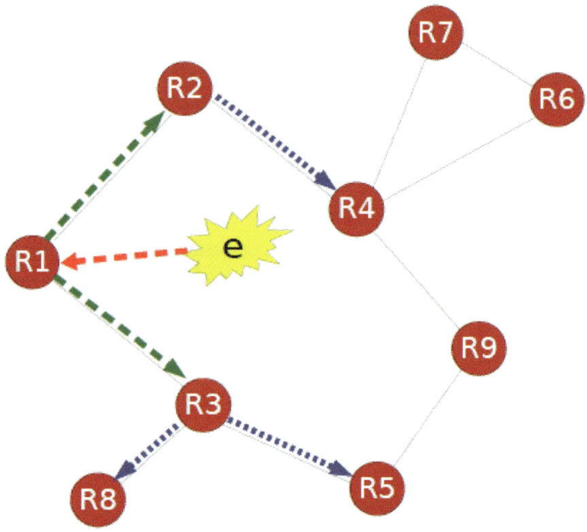

Fig. 3.6 Greedy improvement of auction based protocols.

To illustrate k-SAAPG, let's assume that 1-SAAP protocol is used to select (possibly) the best robot in scenario in Fig. 3.6. According to that protocol, collecting robot R1 would decide that the winner is robot R2 (or R3 but not both) since it is the closest among the three robots R1, R2 and R3. However, there is a closer robot (R4) that can be found by applying greedy phase. Localized additional 1-SAAP auctions initiated by R2 or R3 are illustrated by blue dotted lines. Note that there will be two greedy rounds of auctions. First round will be initiated by R2 and the second by R4, which will not find any better robot.

Another idea that features good communication overhead reduction is when monitored area is configured in cells with sensors randomly placed within cells and every cell has one static node called agent (see Fig. 3.7). There are two possible scenarios. First, the sensor routes to the agent in its cell informing it about event. If there is no idle robot in that cell, the agent searches for near available robot through column/row of agents or using some other search method. Alternatively, the sensor could route to robot as depicted on Fig. 3.7. It is shown that routing directly to agents is much better than routing to robots in terms of lower communication overhead [9].

Another application of auctions is to improve the basic iMesh protocol that is reported to have around 95% efficiency in finding the closest robot

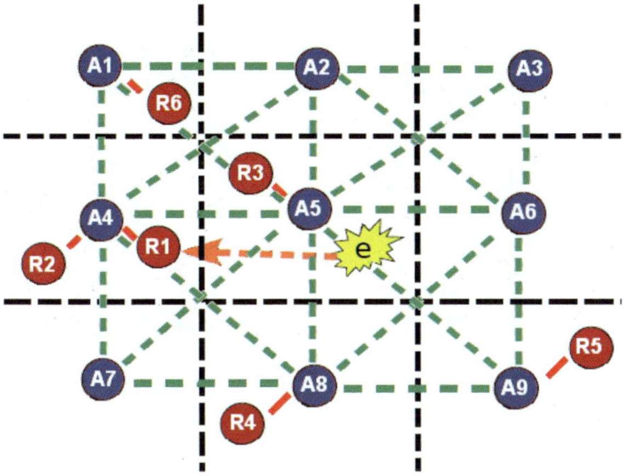

Fig. 3.7 Architecture with agents and robots (sensors are not shown).

in grid networks [14]. Illustration of cross-lookup and scenario when iMesh doesnt find the closest robot is already given in Fig. 3.4. We proposed further improvement of iMesh algorithm using auctions (i.e., k-SAAP) for dense networks with random deployment of service providers [18]. It is shown that efficiency is increased to 99.3% by means of auctions. However, both protocols do not guarantee the closest service discovery in all scenarios, especially for sparse networks or in the presence of holes in the network.

To overcome the problem of void areas in the sensor network that may impose various routing or service discovery problems, we propose a novel localized querying scheme for robot discovery based on Voronoi cells. The scheme is depicted in Fig. 3.8. The basic idea is that every robot in the network locally floods its neighborhood until it reaches an area where other sensors already have information about other robots. Robots in the crossing area stop further dissemination of location data. This area is actually the Voronoi cell border (green lines in Fig. 3.8). In such a way, every sensor has the information about the closest robot and service discovery is not needed. It can be shown that proposed scheme works well with void areas in the sensor network. C-shaped void area and appropriate network graph with 5 robots is depicted on Fig. 3.8(b). In addition, the proposed scheme has good robot location update property [15].

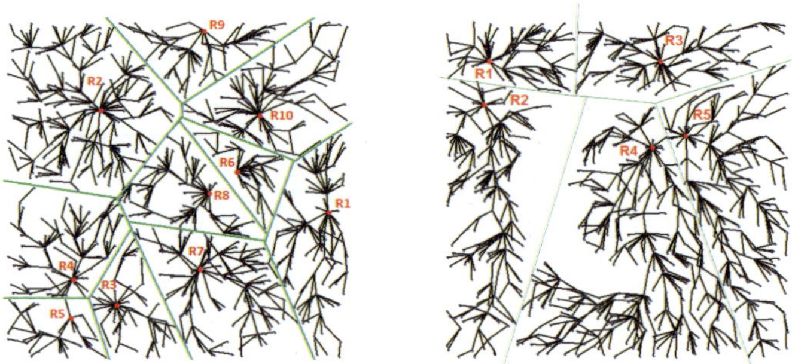

(a) 8 robots

(b) C shape - 5 robots

Fig. 3.8 Voronoi cell based robot location dissemination for random-uniform (with 8 robots) and network with C-shaped void area (with 5 robots).

3.3 Multiple-event Scenarios

In the case of multiple-event scenarios when communication costs are not included, several different scenarios can be explored. Robots can be assigned with single tasks or with multiple consecutive tasks. Events can occur concurrently or non-concurrently (sequentially). There can be tasks that need more than one robot. If events occur non-concurrently and time between occurrences of two events is long enough, this is the case of sequence of single events (iterated assignments) and can be solved as discussed in the previous section. If all events occur concurrently, we have several different assignment cases.

First scenario when only one robot is assigned to one task is called single task assignment-single robot task (STA-SRT). Multiple task assignment is allowed in multiple task assignment-single robot task (MTA-SRT) scenario. If there are tasks that need more than one robot but robots can be assigned with only one task, this scenario is called single task assignment-multiple robot task (STA-MRT). The last scenario is called multiple task assignment-multiple robot task (MTA-MRT) in which tasks that need more than one robot exists and robots can be assigned with more than one task. These four scenarios are depicted in Fig. 3.9 for the case of four robots (R1-R4) and four tasks (T1-T4).

There are a lot of solutions for the STA-SRT problem. Most of them assume that robot communication graph is complete. Solutions are often a

	A. STA–SRT				B. STA–SRT				C. STA–MRT				D. MTA–MRT			
	T1	T2	T3	T4	T1	T2	T3	T4	T1	T2	T3	T4	T1	T2	T3	T4
R1	1				1	1	1		1		?	?	1	1		
R2			1						1		?	?	1			
R3		1						1		1	?	?			1	
R4				1						1	?	?			1	1

Fig. 3.9 Different scenarios for multi-event assignments.

variant of integer linear programming problem or assignment problem [12] (e.g., optimal assignment problem [11], classical assignment problem [25], linear sum assignment problem [4]). The main goal is to find such assignment that minimizes overall traveled distance by robots thus minimizing energy consumption spent on mobility. The problem can be formalized as follows

$$\text{Find:} \qquad T = \{t_{ij}\} - \text{assignment matrix} \tag{3.1}$$
$$i = 1, ..., N_R, j = 1, ..., N_E$$
$$N_E = \text{number of events}, N_R = \text{number of robots}$$

$$\text{Minimize:} \quad C_{min} = \sum\sum m\left[\tfrac{1}{\text{meter}}\right] \times d_{ij}[m] \times t_{ij} \tag{3.2}$$
$$m = \text{movement cost per meter}$$
$$d_{ij} = \text{distance between event } j \text{ and robot } i$$

$$\text{Constraints: } \textstyle\sum_{i=1}^{N_R} t_{ij} = 1, \forall j \tag{3.3}$$
$$\textstyle\sum_{j=1}^{N_E} t_{ij} = 1, \forall i \tag{3.4}$$

If energy or time is an issue that is needed to be taken into consideration, previous problem can be expanded to an instance of assignment problem with side constraints [25], by adding additional constraints that take energy or time into account as follows

$$\sum_{j=1}^{N_E} E\left(T_{ij}\right) \times t_{ij} \leq E_l\left(A_i\right); i = 1, 2, \ldots N_R \tag{3.5}$$
$$\text{where } E\left(T_{ij}\right) = m \times d\left(T_{ij}, T_{i(j-1)}\right). \tag{3.6}$$

Total spent energy must be within energy budget (E_I) for every robot and is given as constraint (3.5). Energy spent for performing task consists of energy needed for movement to come to action area. This is given by equation (3.6). If energy for performing action is not constant (e.g., there are different types of tasks) it can be added in equation (3.6). Introducing energy constraint (as a relevant resource in WSRNs) in model seems to be well justified at first glance. However in practical examples, costs and spent energy per task are highly correlated values and it should be proved that this model expansion is well justified.

Time bound requirements that are common for real time applications are not included so far. If it is important, time needed to complete an action could be included as a simple side constraint:

$$t_M\left(T_{ij}\right) \leq t_r \tag{3.7}$$

where t_M is time needed to come to action area (can be calculated if robot speed is known) and t_r is response constant constraint. Constraint (3.7) can be expanded with the complete response and action time constraint (t_T):

$$\sum_{i=1}^{N_R} \sum_{j=1}^{N_E} \left(t_M\left(T_{ij}\right) + t_r\right) \leq t_0 \tag{3.8}$$

where t_0 is observation time.

In this centralized scenario, it is assumed that everything (all cost, all tasks, all positions and all energies) is known a priori. Communication costs are neglected. Task assignment is in this case NP-hard problem, but a solution can be found by means of some heuristic optimization solver (for example by Hungarian algorithm [13]). In real-time applications it is not possible to know everything a priori, thus making this solution unrealistic for real applications. It can be used for comparison purposes. Another problem that can't be solved with single task assignment is if there are more tasks than robots. If events don't occur concurrently, but rather as one group of events at a time, problem can be solved as an iterated version of the optimal assignment problem.

If all tasks are not known a priori (e.g., occur dynamically or even randomly), the problem formulated in (3.2) cannot be used. A potential heuristic that can be used is an instance of Greedy algorithm. Algorithm would be: When a new task is introduced, assign it to the robot that incurs lowest cost or to best fit robot as in MURDOCH [10] (greedy market-based algorithm). The definition of the best fit depends on metrics being used.

In [22], the authors propose a solution of task assignment problem. The basic assumption is that multiple events occur concurrently, and the number of robots is equal to the number of events. It is a centralized solution which disregards the communication costs. It is assumed that all of the robots and events are aware of the positions of all of the others robots and events. The robots are considered to be men, and the events women. They all create preference lists of the opposite sex based on distance, in such a way that the closer ones are the more preferred. Then, the robots are paired with events by using Gale-Shapley Stable Marriage Algorithm (SM). This algorithm provides matching between sets of men and women in such a way that all of the elements are married, and all the marriages are stable. That means that there is no alternative pairing of man M and woman W such that they both prefer each other more than their current spouses.

The authors of [22] claim that SM heuristic provides the solution that is close to the theoretical brute-force approach which would take every possible matching into consideration. However, our current research shows that this result can be improved at cost of only slight computational overhead. This is due to the fact that stable matching might not be the most energy efficient in terms of total robot movement. Thus, after the execution of stable matching algorithm, we execute Pairwise Distance-based Matching (PDM) algorithm [16] which checks for every two pairs (R1, E1) and (R2, E2) of robots and events which are married if the lines R1E1 and R2E2 intersect. If so, they exchange spouses and form new pairs (R1, E2) and (R2, E1). Total traveling distance is now reduced, because $|R1E2| + |R2E1| < |R1E1| + |R2E2|$, which allows the algorithm to converge towards the solution where no such further exchange is possible (See Fig. 3.10).

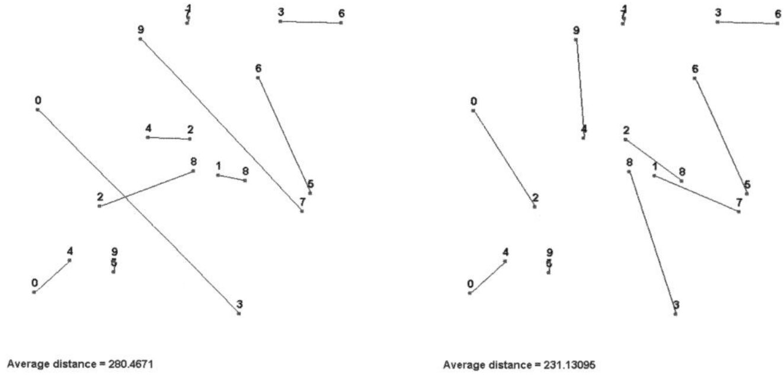

(a) after SM (b) after PDM

Fig. 3.10 Robot-event matching after SM (left) and PDM (right).

In [26], another solution for the same problem is proposed. The assumptions remain the same (equal number of robots and events and global topology knowledge). Another algorithm called Maximum Matching Algorithm (MMA) is introduced, which is designed not to minimize total traveled distance, but to minimize standard deviation between travel distances of individual robots. This leads to a more balanced workload among robots which is important for maximizing the system lifetime in multi-round scenario. However, our research showed that PDM outperforms MMA in this respect. In fact, in multi-round case, we showed that even better results are achieved when we change the metrics that is used in creating preference lists. The new metrics take the remaining energy of robots into account, in such a way that robots with less energy exhibit more resistance to new movements. The resulting Pairwise Resistance Matching (PRM) showed the best performance in terms of total system lifetime, in our simulations of multi-round scenarios [16].

If there are more tasks than robots, each robot should be assigned with multiple consecutive tasks. Intuitively, in cases where robot resources and task priority are not an issue (e.g., unlimited power supply), number of tasks can be divided into w number of subsets. First $(w-1)$ subsets consist of number of tasks equal to the number of robots and these problems are reduced to classical assignment problems. Last set of tasks consists of number of tasks that are less than or equal to number of robots. Hence, this problem is a classical assignment problem or optimal assignment problem. The main problem with this heuristic approach is how to partition events. Thus, actual assignments can be highly suboptimal (in terms of overall distance traveled or needed time).

If robot resources are important and not unlimited, this problem can be recognized as a Generalized Assignment Problem (GAP) [25]. GAP is an example of a one-to-many assignment problem that recognizes resource limits. Optimization problem is to minimize costs incurred by assigning NR robots to N_E tasks, while robots resource is limited and one task is assigned to single robot. However, it assumes that costs are constant. Another heuristic approach, where number of robots and events are equal, is to initially solve the problem as discussed in previous section, and afterwards use Greedy algorithm to assign any remaining tasks when robots become available (this is called ST-SR-TA approximation algorithm in [11]).

In [26], the authors addressed the multiple task assignment problem in the following way. The number of robots is N_R, and the number of events is N_E ($N_R < N_E$). First, the set of events is partitioned into R clusters, following one of three partitioning schemes (k-means, MaxMin or balanced). Then, MMA algorithm previously introduced is applied to the set of robots and the set of clusters, where distance between a robot and a cluster is defined as minimal distance between robot and any of the events in that cluster. When the robots are matched with clusters, they first travel to the closest event in the cluster they are matched to, and then they visit all the other events in the cluster sequentially, by applying an algorithm that solves Traveling Salesman Problem (TSP). We have proposed another algorithm that deals with the same problem called Sequence Dispatch (SQD). It is a simple greedy approach which works as follows. First, a robot R from the set of robots and event E from the set of events are chosen, such that distance |RE| is minimal. Then, E is excluded from the set of events. R adds E to its visit schedule and updates its current position to the position of E. The procedure described above is repeated, until all of the events are included into visit schedules of robots. Our simulations show that MEM outperforms all of the clustering schemes combined with TSP, proposed in [26], both in single and multi-round scenarios. Again, in such scenarios we showed that even better results are achieved with taking remaining robot energy into account, which led us to the improved version of SQD called Sequence Resistance-based Dispatch (SRD) [16].

How to make task allocation in cases where tasks need more than one robot are reported in [11]. One example of such tasks is heavy box pushing or coordinated exploration. It is assigned as an instance of Set Partitioning Problem SPP (maximum utility version). It is strongly NP-hard but there are several heuristics for SPP. It remains to be seen if SPP heuristics are good enough to be applied to robot task assignment. If robots can be assigned with more than one task and there are tasks that need more than one robot, it is case of multiple task assignment-multiple robot tasks allocation problem. To the best of our knowledge, no heuristic for this difficult problem is reported and this can be an area for future research.

In all previous scenarios communication costs are neglected. In cases when communication cost is included and multiple events occur concurrently, the distributed version of task assignment problem can be decomposed in two phases, first of which is service discovery. This can be achieved by means of quorum, iMesh, or other service discovery algorithm. In the next phase when all the event-sensing nodes have discovered the nearest robot location, they all send service request messages to corresponding robots. During this phase the robots create lists of service requests. The problem is that one robot may receive requests from events that may not be close to each other. Thus, it is necessary to use robot-robot communication in order to allow exchange of service requests between neighboring robots which will lead to more energy efficient dispatch schedules. To achieve this goal, we propose some of the auction-based schemes, introduced in [19]. The resulting localized algorithm (MAD) [16] based on information mesh infrastructure for finding nearby robots, and local auctions in a robot network to perhaps find a better one that outperforms existing solutions based on stable marriage problem, k-means clustering and traveling salesman tour in terms of lower communication costs and longer network lifetime. What remains as a future research challenge is to consider solutions for two classes of hard problems, namely, single task assignment-multiple robot tasks and for multiple task assignment-multiple robot tasks.

References

[1] Akkaya, K., Guneydas, I. and Bicak, A. (2010). Autonomous actor positioning in wireless sensor and actor networks using stable-matching, *Int. J. Parallel, Emergent and Distributed Systems* **25**, 6, p. 439–464.

[2] Akyildiz, I. F. and Kasimoglu, I. (2004). Wireless sensor and actor networks: research challenges, *Ad hoc networks* **2**, 4, pp. 351–367.

[3] Bekey, G. and Yuh, J. (2008). The status of robotics, *IEEE Robotics Automation Magazine* **15**, 1, pp. 80–86.

[4] Burkard, R., Dell'Amico, M. and Martello, S. (2012). Assignment problems, in *Linear Sum Assignment Problem, (SIAM Society for Industrial and Applied Mathematics)*, pp. 73–143.

[5] Chung, H., Oh, S., Shim, D. and Sastry, S. (2011). Toward robotic sensor webs: Algorithms, systems, and experiments, *Proc. of the IEEE* **99**, 9, pp. 1562–1586.

[6] Dias, M., Zlot, R., Kalra, N. and Stentz, A. (2006). Market-based multirobot coordination: A survey and analysis, *Proc. of the IEEE* **94**, 7, pp. 1257–1270.

[7] Falcon, R., Nayak, A. and Stojmenovic, I. (2013). Robot-assisted wireless sensor networks: Recent applications and future challenges, in S. Basagni, M. Conti, S. Giordano and S. I. (eds.), *Mobile Ad Hoc Networking*, pp. 737–768.

[8] Farinelli, A., Iocchi, L. and Nardi, D. (2004). Multirobot systems: a classification focused on coordination, *IEEE Transactions on Systems, Man, and Cybernetics, Part B* **34**, 5.

[9] Gasparovic, B. and Mezei, I. (2011). Auction aggregation protocols for agent-based task assignment in multi-hop wireless sensor and robot networks, in *IEEE/ASME International Conference on Advanced Intelligent Mechatronics (AIM)*, pp. 247–252.

[10] Gerkey, B. and Mataric, M. (2002). Sold!: Auction methods for multi-robot coordination, *IEEE Trans. on Robotics and Automation* **18**, 5, pp. 758–768.

[11] Gerkey, B. and Mataric, M. (2004). A formal analysis and taxonomy of task allocation in multi-robot systems, *Intl. J. of Robotics Research* **23**, 9, pp. 939–954.

[12] Kaleci, B., Parlaktuna, O., Ozkan, M. and Kirlik, G. (2010). Market-based task allocation by using assignment problem, in *IEEE International Conference on Systems Man and Cybernetics (SMC)*, pp. 135–141.

[13] Kuhn, H. (1955). The hungarian method for the assignment problem, *Naval Research Logistics Quarterly* **2**, pp. 83–97.

[14] Li, X., Santoro, N. and Stojmenovic, I. (2009). Localized distance-sensitive service discovery in wireless sensor and actor networks, *IEEE Transactions on Computers* **58**, 9, pp. 1275–1288.

[15] Lukic, M. and Mezei, I. (2013). Localized querying and location update service in arbitrary wireless sensor and robot networks, in *in preparation*.

[16] Lukic, M. and Stojmenovic, I. (2013). Energy-balanced matching and sequence dispatch of robots to events: Pairwise exchanges and sensor assisted robot coordination, in *The Tenth IEEE International Conference on Mobile Ad-hoc and Sensor Systems (MASS)*.

[17] Melodia, T., Pompili, D., Gungor, V. and Akyildiz, I. (2007). Communication and coordination in wireless sensor and actor networks, *IEEE Transactions on Mobile Computing* **6**, 10, p. 1116–1129.

[18] Mezei, I., Lukic, M., Malabasa, V. and Stojmenovic, I. (2013a). Auctions and imesh based task assignment in wireless sensor and actuator networks, *Computer Communications, (Elsevier)* .

[19] Mezei, I., Malabasa, V. and Stojmenovic, I. (2010). Robot to robot: Communication aspects of coordination in robot wireless networks, *IEEE Robotics and Automation Magazine* **17**, 4, pp. 63–69.

[20] Mezei, I., Malabasa, V. and Stojmenovic, I. (2013b). Greedy extension of localized auction based protocols for wireless actuator task assignment, *Ad Hoc and Sensor Wireless Networks* **17**, 1-2, pp. 73–85.

[21] Mosteo, A. and Montano, L. (2010). A survey of multi-robot task allocation, Tech. rep., AMI-009-10-TEC.

[22] Nayak, A. and Stojmenovic, I. (2010). *Wireless Sensor and Actuator Networks: Algorithms and Protocols for Scalable Coordination and Data Communication* (Wiley Blackwell).

[23] Osunmakinde, I. (2012). Multi-robot coordination, in R. J. et al. (ed.), *Mobile Intelligent Autonomous Systems*, p. 271–280.

[24] Parker, L. (2008). Multiple mobile robot systems, in S. Bruno and K. Oussama (eds.), *Springer Handbook of Robotics*, pp. 921–941.

[25] Pentico, D. (2007). Assignment problems: A golden anniversary survey, *European Journal of Operational Research* **176**, p. 774–793.

[26] Wang, Y., Peng, W. and Tseng, Y. (2010). Energy-balanced dispatch of mobile sensors in a hybrid wireless sensor networks, *IEEE Transactions on Parallel and Distributed Systems* **21**, 12, pp. 1836–1850.

[27] Yao, Z., Dai, X. and Ge, H. (2012). Quantitative and qualitative coordination for multi-robot systems. *Artificial Intelligence and Computational Intelligence, LNCS 7530,* , pp. 755–761.

Chapter 4

Mobile Robot Deployment in the Context of WSN

Milan Erdelj and Karen Miranda

FUN Research Group, Inria Lille – Nord Europe, France

Abstract. *The advances in mobile robotics allow us today to add the mobility concept into many different classes of wireless sensor networks (WSN) or wireless sensor and actuator networks (WSAN) applications. The deployment of mobile sensors is possible and useful in many application scenarios, ranging from the environmental monitoring, e.g., volcano activity, dispersion of fire, pollutants or gas plumes, and public safety applications (event or object surveillance), to the industry (structure and machinery health) and military applications (automated warfare, land mine detection). Manual sensor deployment represents a rather difficult task to achieve in such type of applications due to various reasons that will be discussed in this chapter. The use of mobile agents, i.e., robotic platforms equipped with sensory and motion capabilities, allows us to overcome these difficulties by deploying the sensor network in a random manner and applying the self-repositioning of self-deploying techniques.*

4.1 Notions of Mobile Robot Deployment

4.1.1 *Sensor network deployment*

The notions of *deployment* and *deployment objective* are hard to define since they depend on the actual application of the WSN/WSAN. Furthermore, the concept of *deployment quality* strongly depends on the deployment goals, sensor, and environment characteristics. Indeed, different deployment solutions can be envisaged in the case of sensors with limited communication and movement capabilities, or the absence of knowledge regarding the deployment environment. In other words, the quality of the deployment is not comparable in the case of mobile sensors with total knowledge of the environment and the availability of the absolute localization techniques, than in the case of absence of any localization technique followed by the completely unknown deployment environment. Bearing in mind that the majority of the applications focuses on a certain type of event (or a set of events) monitoring and data acquisition, the deployment can be referred to as the process of optimally placing a group of sensors (static and/or mobile) in an environment containing the events of interest. In the context of environmental monitoring, the deployment quality will notably depend on the environment covered area, deployment speed, and energy consumption, just to name a few.

In general, the sensors are usually deployed in a random or deterministic manner. Former are hardly feasible in any other situation than the small network deployed in the known environment. The necessity of larger sensor networks in an unknown environment leaves us with the random deployment as the only choice. The random deployments in an unknown environment is usually done by scattering the sensors over an area of interest, such as volcano or forest, from an aircraft. As expected, certain number of sensors deployed in such a manner will not be usable due to failures caused by the aircraft scattering. In order to guarantee the quality of such a deployed network (notably regarding the covered area), the number of sensors deployed must be greatly larger than the optimal number, which increases the overall costs of the network.

4.1.2 *Sensor mobility*

A way to improve the deployment quality in terms of coverage is to introduce the mobility capabilities into the network. In the case of an initial random deployment, static nodes could be replaced with mobile substitutes, which could increase the cost of the network, however, the deployment quality would increase as well. Another method of introducing mobility into the WSN is the addition of a few mobile robots (not necessarily with sensing capabilities) that are used to displace the static sensors, thus increasing the deployment quality.

Including the robot mobility in the WSN deployments allows us the following:

- the possibility to resolve problems that could appear in the network that are not solvable by static nodes,

- increasing the network robustness by automated sensor node replacements,
- the adaptability to unknown or dynamic environments,
- increasing the speed of data processing and routing through the use of parallelism.

There are three major types of mobility that are considered in the context of WSN/WSAN:

(1) *Static (immobile) WSN.* In this first case, the sensors in the network do not possess any kind of locomotion or displacement capability. Furthermore, there are no entities that could interact with or displace sensors in the network. This type of WSN is the most widespread and used in most applications worldwide.
(2) *Assisted mobility.* This second type of mobility assumes a sensor network composed of static sensors that are unable to move autonomously. However, these sensors are usually mounted on different types of mobile agents that provide them with mobility. These moving agents depend on the specific application and their moving pattern is not controllable, however it can be modeled in a certain way. Examples of this type of mobility are the sensors mounted on vehicles, animals, or people [10, 14].
(3) *Controlled mobility.* Finally, the third type of sensors mobility assumes that the network is entirely or partly composed of mobile sensors (mobile robots) that can be manually or self-controlled. This type of mobility allows us to increase the deployment quality in a way that suits the best to the user of the network. Controlled mobility has received much attention in recent years due to the ever expanding possibilities for different applications that were not possible beforehand (notably area exploration and rescue missions).

In this chapter, we focus our attention on the controlled mobility, i.e., the third type of mobility, and we will refer to it in the remainder of the chapter simply as *mobility*.

4.1.3 *Deployment of multi-robot systems in the context of WSN*

By introducing more that one mobile agent with or without sensing capabilities in the sensor network, the WSN may be observed from the point of view of multi-robot systems (MRS). MRS are nowadays used in wide variety of cases, including the scientific, environmental, industrial, and military applications. All of these applications require the high deployment quality in the means of covered area, speed, and energy used which is generally achieved thanks to the carefully chosen controlled mobility technique, which still represents a challenging task.

All the approaches to sensor deployment that include controlled mobility can be classified into two deployment schemes: centralized and distributed deployments. The centralized approach assumes the existence of a central entity that is not necessarily a part of the set of mobile sensors. The role of the central entity in this type of deployment techniques is to collect all the necessary information about all the sensors in the network and the deployment environment itself, process this information regarding the goal of the deployment, choose the optimal positions for each sensor in the network, and finally, direct each individual sensor towards its future destination. This type of approach can achieve excellent results in the static environment, since the optimization algorithms can be applied in order to achieve the optimal deployment. However, the necessity of the global network information acquisition imposes high computational cost in energy, time, and storage space, that collides with the concept of WSN composed of cheap sensors with limited processing power. Furthermore, due to the centralized approach depending on central entity, the complete network is dependent on the errors and failures that can happen in the central entity, which makes the network highly vulnerable. Finally, the scalability of the network in this case represents another huge problem, since the central entity has to manage ever increasing amount of information in real time. The aforementioned problems, together with the dynamics in the most practical environments that increase the complexity of the computation and communication, make the centralized approach infeasible in most practical applications.

On the other hand, the distributed approach easily copes with the problems of the dynamic and unpredictable environments, as well as the problem of scalability, by allowing each robot to calculate its own behavior and mobility pattern depending on the perceived local neighborhood and environment information. In this manner, the computation complexity is reduced to a limited set of locally perceivable neighboring sensors, whatever the size of the complete network is. The goal of the distributed deployment techniques is to combine all the movement decisions that are brought locally and combine them in order to approach the optimal solution achievable by the centralized approach. The drawback of the distributed approach is that the lack of complete knowledge makes impossible to achieve the optimality. However, bearing in mind the environment conditions in practice, followed by the absence of scalability and computational complexity issues, in this chapter we focus our attention only on the distributed approaches to multi-robot WSN deployments.

4.1.4 *Network connectivity problem*

Maintaining the connectivity among the sensors in the network is one of the essential tasks during the deployment [1, 4, 21]. By assuming the distributed approach to the deployment, and thus assuming the localized and constrained knowledge of the network, it is impossible to achieve the optimal (or locally optimal) deployment if the sensors in the network do not form a connected graph. Therefore, it is usually assumed that the set of sensors form a connected network, with tendency to preserve the connectivity at bootstraps all along the deployment procedure and the network lifetime.

In the context of the communication graph that represents the WSN, the connectivity maintenance problem is essentially the problem of moving the sensors in such a way to avoid disconnections in their communication graph. The basic idea that underlies every connectivity preservation technique is to restrain the movements of each mobile sensor depending on the corresponding communication graph. In the general case, the introduced constraints are based on the one-hop communication link and they continuously depend on the sensors' positions. The optimal connectivity preservation technique could be defined as the technique where movement constraints are "loose" enough to minimally constrain the sensor motion.

4.1.5 *Generalized robot deployment algorithm*

All the different deployment approaches that we examine in this chapter and that depend on the specific application of WSN, can be described with one general algorithm with a rather simple structure (Algorithm 4.1). As we adopted the distributed nature of the deployment, the algorithm is iterative and comprises three essential parts: neighborhood discovery, movement target computation, and movement towards the computed target point. In the neighborhood discovery part, the robot transmits its own position and receives the positions of neighboring robots in the deployment field. This information is used in order to construct the communication and sensing graph based on some graph reduction technique, for example, Minimal Spanning Tree, Gabriel Graph, and Relative Neighborhood Graph. The second part of the algorithm employs different probabilistic or geometrical techniques to choose the best potential displacement target point while applying the connectivity preservation constraints. Finally, the third part executes the movement towards the selected point. Further details about each part of the generalized deployment algorithm depend on the sensor type, environment characteristics, and the application specifications.

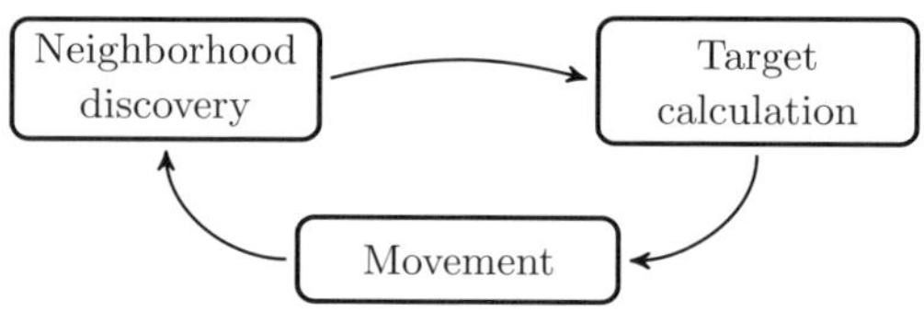

Fig. 4.1 Generalized robot deployment algorithm.

In the remainder of the chapter, we discuss the coverage problem in the WSN (Section 4.2), sensor deployment approaches (Section 4.3) and techniques (Section 4.4), followed by the discussion on different types of mobile robots in the context of WSN (Section 4.5). We discuss some open issues regarding the robot deployment in the Section 4.6 and conclude the chapter in Section 4.7.

4.2 Coverage Problem

The main task in WSNs is to monitor a given target and to transmit cooperatively the collected data over the network to a main location, therefore, the nodes must be able to capture every event about the specific target. Hence, how well a target is sensed by a WSN is an issue called coverage. A target is covered when it is within the sensing range of at least one sensor. The quality of the network is attached to the coverage degree and is often used as a performance metric as well.

The coverage requirements depend directly on each particular coverage problem [3, 17]. In general, the coverage problems are classified according to their ultimate goal into three categories: full coverage, barrier coverage, and sweep coverage.

4.2.1 *The full coverage problem*

The full coverage problem is often known as blanket coverage or region coverage problem [2, 15]. This problem arises when the objective is to cover a whole field of interest (FoI), i.e., every point within the field must be sensed by at least one node. In Figure 4.2(a), we illustrate the full coverage problem. One example of WSN applications presenting a full coverage problem is in a vineyard, where we have a big field and the sensors must collect information about, for example, temperature or humidity.

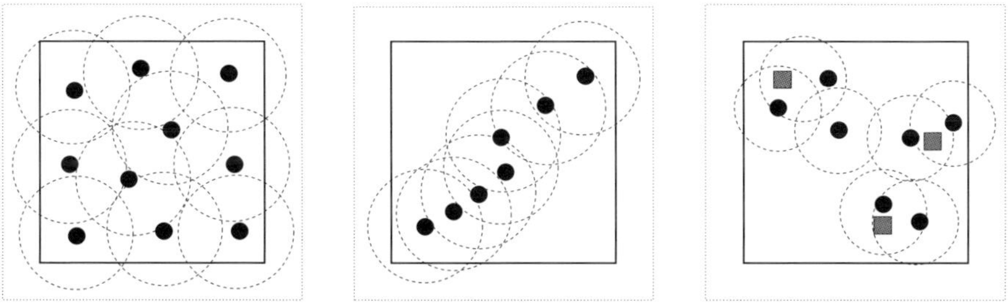

(a) Blanket coverage (b) Barrier coverage (c) Sweep coverage

Fig. 4.2 Robot deployment coverage problems.

4.2.2 *The barrier coverage problem*

The concept of barrier coverage differs substantially from the concept of full coverage [7]. The goal of full coverage is to deploy a set of sensors to cover an entire zone meanwhile the border coverage aims to detect intrusions in or across the target zone. Therefore, the sensors are deployed only through the crossing paths of FoI in order to reduce the possibility of a given intruder crossing the field undetected. Figure 4.2(b) represents an example of barrier coverage, where the nodes are spread over a line. A key example is a border line, where there is a huge effort to detect any crossing violation. For this example, the sensor should be deployed in a line to surround the field of interest. In particular, it is an interesting proposal to work jointly with sensors and cameras to detect any unauthorized crossing.

4.2.3 *The sweep coverage problem*

In general, when the sensors are deployed to solve the FoI or barrier problems, they remain static at their positions [9]. However, the coverage strategy changes when the network must cover only a few Points of Interest (PoI) or targets. These specific PoI may be static or dynamic in the time, i.e., the points that are of interest at a given time can change dynamically after a while. This means that the sensors must be redeployed in order to cover the new PoI. Therefore, the concepts of robotics are very important in the context of sweep coverage to provide the sensors with mobility capabilities. Figure 4.2(c) illustrates a field with several PoI inside, the sensors are deployed to cover a few specific points leaving the rest of the field uncovered.

4.3 Deployment Approaches

4.3.1 *Deterministic deployment*

The sensors' positions are set up before the actual deployment following a predefined shape, as a diamond for example. Taking the application into account, the sensors are deployed uniformly over the target region or with a weighted distribution to cover a certain number of targets. An example of the deterministic approach is a grid-based deployment where the nodes are equidistantly separated according to a grid shape. Figure 4.3(a) depicts a deterministic deployment, the nodes are evenly placed over the field.

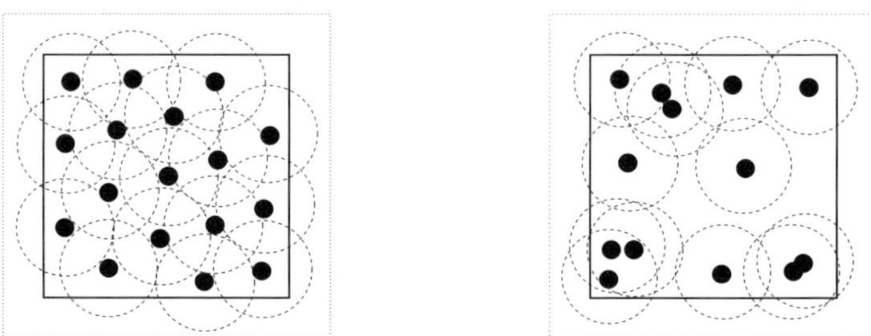

(a) Deterministic (b) Random

Fig. 4.3 Robot deployment approaches.

4.3.2 *Random deployment*

A deterministic deployment for many applications is unpractical or simply not possible to do. In these cases, the sensors can be deployed randomly, e.g., dropping the sensors from an airplane. Hence, the localization of the nodes is unknown before the deployment. In general, it is possible to use a stochastic distribution, such as uniform, Gaussian, or Poisson to model the sensor distribution. Figure 4.3(b) depicts a random deployment, where the nodes are randomly placed over the field.

4.3.3 *Static*

Once the sensors are placed, they remain at the same position. Generally, the deployment algorithms consider static nodes since they are cheaper than mobile ones. The static deployment may be at the same time random or deterministic.

4.3.4 *Dynamic*

A dynamic deployment assumes that some or all sensors have mobility capabilities. Mobility allows the deployment of nodes on-demand, controlling the movement to obtain particular network topologies, specially in unknown environments. This scheme, also, represents the possibility of *re-deploy* the network in case that the environment's conditions change [20].

Usually, the deployment algorithms proposed in the literature consider a subset of the requirements presented above for particular scenarios. For example, in case of blanket coverage of a building it is easy to envision the usage of a deterministic algorithm. Such an algorithm should optimize the number of nodes, the distance between them, and the coverage range. Conditions change when there is no any a priori knowledge about the environment, for example within forest. In such a case, some areas of the field could be not covered at all, several nodes could be too close or too far from each others. Hence, the random topology may not fulfill the coverage and connectivity requirements. In this latter case, it is possible to use mobile nodes to redeploy them and then guarantee the coverage.

4.4 Field Coverage Optimization

The result of the deployment phase has close implications with the algorithms used to optimize the coverage. Therefore, the algorithm design must consider, both coverage and deployment in order to optimize the network performance [18].

4.4.1 *The pattern-based technique*

In order to localize the PoI, the sensors are deployed in a predefined regular pattern, such as spiral, hexagon, circular, or triangle. Once the sensors are deployed following the pattern, their final positions are computed either by considering the global coverage or by using a mobile sink sensor that acts as a facilitator to help other sensors to find their final position. A main drawback of this type of techniques is that they do not provide connectivity during the whole deployment procedure. In Figure 4.4(a), we present an example of pattern-based coverage, where the nodes are deployed following a triangle form.

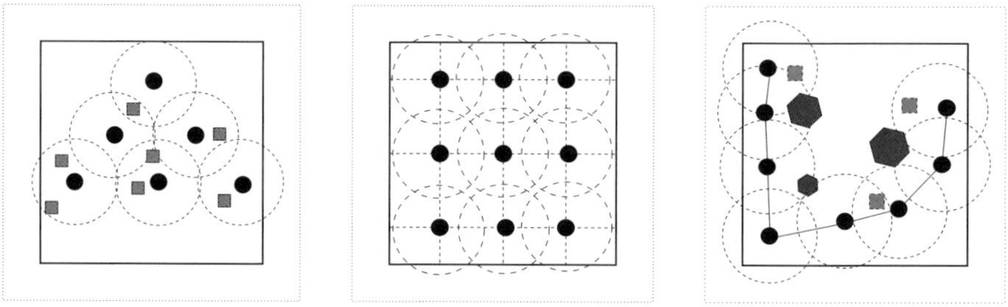

(a) Pattern-based (b) Grid-based (c) Virtual force-based

Fig. 4.4 Robot deployment techniques.

4.4.2 *The grid quorum-based technique*

The field is divided into several small grids forming cells. Each cell has a load depending on the number of sensors inside the cell. Therefore, the coverage and connectivity are a function of the grid's size. Thus, the sensors must move in order to balance the cells' load. Figure 4.4(b) depicts the grid quorum coverage, all the nodes are even placed all across the field creating a well-defined grid.

4.4.3 *The virtual force-based technique*

The virtual force technique is based on the concept of electromagnetic particles, where the particles attract or repel each other according to the particles characteristics. Mapping the concept to sensor networks, the preferential coverage areas work as attraction forces, meanwhile, the obstacles work as repulsion forces. Hence, these forces are computed based on the sensor's neighborhood to allow the computation of the sensor's next movement. Figure 4.4(c) illustrates a field with several PoI inside, the sensors are deployed avoiding the obstacles and preserving the connection with the sink node.

Special algorithms are required after the deployment to guarantee area coverage and connectivity in the network. Such algorithms, centralized or distributed, manage problems, such as, connectivity preservation, saving energy, and range adjustment that help to optimize the network performance.

4.5 Mobile Robots in the Context of WSN

In this section, we introduce different types of mobile robots that could be used as an integral part of the wireless sensor network for a specific application and depending on the deployment area characteristics. Furthermore, we discuss some problems of robot and sensor interactions withing the network, followed by the set of typical robotic sensor network applications.

4.5.1 *Mobile robots as autonomous vehicles*

Despite being out of the scope of this chapter, it is worth noting that the term *mobile* robot could represent any type of robot with movable parts that is capable of changing its position in a certain way. That notion includes the wide range of industrial robots used in production lines. However, in the context of WSN, we focus our attention only on mobile robots that represent autonomous vehicles whose movements are not limited by their physical size. Hence, mobile robots as autonomous vehicles can be used to explore unknown environments and perform a variety of functions that would normally been performed by humans. They are classified in three large groups depending on their operating environment,

(1) *Ground vehicles (land-based robots).* This is the largest group of mobile robots. It is widely used in different sorts of applications due to its relative simplicity of construction and intuitive operating mode.
(2) *Aerial vehicles (flying drones).* The development of this group of robots is under expansion, notably due to the specific characteristics of the deployment medium. The use of flying drones allows us to avoid the problems of physical obstacles in the deployment field simply by flying over them, backed up by the increased speed of deployment in compari-

son with the land-based robots. This type of mobile robots is often used for the applications of area surveillance and target detection/tracking.

(3) *Marine vehicles (aquatic robots).* This group of robots introduces a specific set of challenges due to the deployment environment characteristics. Being deployed in the water, these robots face two problems: the problem of communication in the aquatic medium and the problem of localization since the global localization techniques are not available if the robot is submerged under water. A significant amount of research is being done in recent years in order to find a satisfying technique of communication, via sound waves rather that the electromagnetic waves, and localization, landmark based instead of global positioning, (see also Chapter 9).

4.5.2 *Mobile robots and the interaction with WSN*

We have already mentioned that the random deployment of static WSN requires a number of sensors that is greater than optimal, which impacts the overall deployment cost. One of the solutions to this problem is the conjunction of a classic static WSN with a set of mobile nodes [8]. In this context, the role of mobile robots is twofold. First, the set of mobile robots serves as mobility provision agents. In this case, the goal is to physically displace already deployed static sensors in the deployment field and thus increase the deployment quality. However, it cannot be guaranteed that in every WSN application, such approach would improve the quality of the deployment while minimizing the deployment costs. As stated before, this problem is strongly dependent on the specific application specifications and, above all, on the deployment environment characteristics. An example of infeasible interaction of mobile robots with static WSN is the WSN for the seismic activity monitoring that is deployed on the ocean floor. Such hostile deployment environment rules out the implementation of a multi-robot system capable of providing the sensors with mobility. In such and similar cases, it is worth considering the trade off between the cost of introducing the mobility versus the additional set of static nodes with the accent put on the improved data acquisition protocol that could achieve better results and increase deployment quality [5].

The second role of the mobile robots in the interaction with the WSN is automated sensor network servicing. Although not directly involved in sensing and acquisition tasks, a set of mobile robots can influence the deployment quality by replacing damaged or discharged sensors with working replacements, or even behaving as a mobile recharging station, thus prolonging the lifetime of the network [12, 16, 19]. The sensor node servicing by a group of mobile robots is a complex problem and may be observed as one deployment problem within another – the application of the WSN is the environmental information acquisition, while the application of the set of servicing robots is both the acquisition of the information regarding the sensor network and the servicing decision and schedule problem.

4.5.3 *Applications of mobile robotic networks*

Generally speaking, there are two classes of applications in which the use of autonomous mobile robots is needed:

- inaccessible or unknown deployment environment inspection (warfare field, structural health monitoring, and machinery inspection [13]),
- and the hazardous environment where the presence of humans can be endangered (minefields, toxic gas leaks, and pollutions source detection [11]).

Regarding the first class of applications, the mobile multi-robotic networks play an important role in the field of electronic and visual reconnaissance, deployment field surveillance, target detection, and identification. Most of these applications have a military connotation up to a certain degree, which is understandable since the huge amount of resources are allocated in order to improve the quality and usability of mobile robotic networks. Military applications focusing on security are present in the second class of applications as well, with the applications such as minefield exploration and mine detection, together with chemical, biological, radiological, nuclear, and explosive reconnaissance problems that need to be solved with the help of mobile robotic networks.

One typical example of the robot deployment for such means is presented in [6]. A set of autonomous mobile robotic vehicles is deployed in the building that suffered an attack and therefore represented an unknown environment with unknown number and placement of people inside. The goal given to the set of robots was to explore the ruined building, locate the people inside and to provide the rescuers with the exact information regarding the situation in the building. Needless to say that the speed of the deployment was of the essence and that all the robots had to collaborate in order to save human lives. This example shows that it is possible to achieve fast and reliable autonomous robot deployments in order to tackle the problem that could not be solved in a different way.

4.6 Discussion and Open Issues

In this section, we discuss some of the common issues in the robotic networks and point out important properties of wireless robotics that should be kept in mind in order to achieve successful mobile robot deployments.

4.6.1 *Communication*

The most important issue in the wireless network is the communication aspect that in this case has its own special characteristics. We can consider two different communication paradigms in this case: direct and indirect communications. The first way of communication is the explicit one, two sensors represented as the sensor nodes in the network can communicate with each other through an established one or multi-hop wireless links. The specific purpose and architecture of a WSN force the sensors to communicate with their neighbors usually only by means of exchanging their position information and transmitting the sensed information towards the data center. However, without loosing of generality, it can be assumed that two sensors can communicate directly via established links.

A second way of communicating is the indirect one – the communication through the signs in the environment (stigmergy). This way of communication is bio-inspired, where instead of creating a direct wireless link, there is no direct communication link that is established, rather the sensors change the deployment environment (by leaving signs, pheromones, etc.) in the way that will be understood by other sensors in the deployment. An example of indirect communication in the set of servicing robots is the communication through messages left at the serviced nodes that can be read by other robots that will pass by. It is worth noting that in this way of communication, there is no need for constant connectivity maintenance, therefore, this type of communication techniques can be used in sparse networks. On the other hand, indirect communication introduces high latency in the network and is not suitable for deployments that require a fast response time.

Another essential problem that arises in the implementation of the robotic wireless networks is due to the wireless channel properties. The wireless communication medium does not have a predictable behavior as the wired channel. This makes the signal strength and the propagation delay highly dependent on the robot hardware, network topology and the properties of the environments (propagation medium, obstacles, etc.). Furthermore, the wireless medium is a broadcast medium, which means that all the nodes in the transmission range of a node transmitting a message, can receive that message, and this poses a problem of unnecessary energy depletion caused by the messages that will be discarded.

In the context of geometry-based deployments, missing short and unexpected long links are common problems, hence, an important issue is the physical link length. The effect of a wireless channel in certain cases makes impossible for two physically close sensors to establish a wireless link. Likewise, in some cases the link can be established even if the distance between two sensors is way beyond the expected maximal communication range. Therefore, the robot deployment becomes a highly complex task if the communication medium properties are taken into account.

These issues, combined with the dynamics introduced in the sensor network that make the network topology change rapidly and unexpectedly, highly affect the availability of communication paths and the quality of the communication.

4.6.2 *Infrastructure based problems*

An integral part of the sensor deployment in most applications is the establishment of the data acquisition infrastructure. The dynamic nature of the robot deployment changes this paradigm in a sense that the network infrastructure must be auto-adaptable to environment conditions. Examples for this are the disaster areas where it is not possible to elect a set of sensors that will play the role of the communication backbone due to the possibility of sensor failures. The complete network should rather be equipped with the mechanisms of overcoming these types of unexpected environment behavior.

Setting up the network infrastructure may seem to be not a so complex task, however, the problems of cost and time to set it up can arise. The cost of the auto-adaptable network infrastructure becomes an obstacle in remote and large construction sites where the robotic network is used for a structure and machinery health monitoring. In most military applications that require fast and reliable response to environmental changes, the network infrastructure reaction time represents one of the major issues.

4.6.3 *Robot robustness, heterogeneity and scalability*

Another major obstacle in the widespread robot deployments is the reliability of mobile robots in the presence of environmental disasters. Robot failures lead to the loss of gathered information and possible network disconnections. These problems could be overcome with the appropriate information routing techniques, however, they do not guarantee that the network will be able to overcome all the problems. In the practical implementations of robotic sensor networks, robots that are used are often heterogeneous, and therefore, prone to different sets of environmental hazards. Information routing protocols usually assume a heterogeneous network of robots, which is not the case in practice. Furthermore, robot robustness is usually examined on the scale of individual robot. In the practical implementations, failures that appear are usually linked to more than one robot that operates in the desired environment, and these failures are sometimes environmentally provoked, meanwhile some other times they are induced by the interaction between the robots. In any case, they are not trivial to detect and overcome.

Standard WSN data acquisition techniques assume a dense sensor network which is still not the case in robotics. The greatest obstacle to achieve in a dense robotic network is the robotic unit price – robotic sensors cannot be considered as cheap sensing devices with limited storage, energy, and processing power (which is a usual assumption in WSN). Due to the sparsity, individual node failures can lead to greater disasters in the networks that could be expected with reasoning inherited from WSN principles.

4.6.4 *Robots, system and sensing model design*

Robot network design generally aims at finding the balance between the simplicity of the individual robotic sensor units and the complexity of the final system that comprises networked robots, but the communication and control flow as well. A number of problems arise due to the lack of understanding of the final application goals and needs, along with the compromise between the highly specialized and generalized modular components used in the construction of the robotic sensors. Modular and reusable components generally reduce the effort and work needed to conceive and implement mobile robots, and in this manner reduce the development costs linked to new component testing. On the contrary, specialized components used in the construction of mobile robots provide the sensor network with the increased suitability and higher performance in the desired application.

Although most of the literature on coverage and connectivity using sensor nodes assume the probabilistic or disk sensing model, the practical implementation issues show that the considered models largely deviate from the reality. Indeed, the practical deployments are environment-dependent and thus cannot be modeled without the detailed knowledge about the deployment characteristics (including the application goal and the environment properties).

4.6.5 *Testing*

The final and fundamental component of any system integration is testing. Robotic networks dedicated for information acquisition applications in the context of WSN, suffer from the same problem that strikes any product in development – the compromise between thorough testing and the necessity to move a designed system to the market quickly. Full system testing is impossible to achieve, above all in the design of mobile robots dedicated for the aforementioned applications, since it is impossible to envisage all the possible situations and hazards that could appear in the real world.

First level of testing is the testing of the used components in the construction of the robotic platform in order to verify their functionality as stated in their specification. When the complete deployment system is integrated, the next level of testing focuses on the functionality of the system itself. This testing phase can take a long period of time in order to ensure the reliability and robustness of the single components integrated in the complex system that is required to fulfill its goals over a long period of time and in various conditions. The last and the critical part of the reliable robotic network is the implementation of the internal self-monitoring techniques that will allow the system itself as well as the individual robots in order to detect, recognize, and solve a set of potential problems that may arise in a real world implementation.

4.7 Conclusion

In this chapter, we discussed the problem of mobile robot deployment by introducing the concept of deployment itself, sensor mobility, and robotic networks. We provided an analysis of different types of deployment techniques and approaches, followed by an analysis of different mobile robot types, the concept of mobile robots in the context of wireless sensor networks, and the set of possible applications for robot deployments. Finally, we concluded the chapter with a discussion on a number of open issues that arise in the real applications and that the robotic sensor network must cope with.

References

[1] Abbasi, A., Younis, M. and Akkaya, K. (2009). Movement-Assisted Connectivity Restoration in Wireless Sensor and Actor Networks, *IEEE Transactions on Parallel and Distributed Systems* **20**, 9, pp. 1366–1379.

[2] Bartolini, N., Calamoneri, T., Fusco, E. G., Massini, A. and Silvestri, S. (2009). Push&pull: autonomous deployment of mobile sensors for a complete coverage, *Wireless Networks* .

[3] Fan, G. and Jin, S. (2010). Coverage problem in wireless sensor network: A survey, *Journal of Networks* **5**, pp. 1033–1040.

[4] Ghosh, A. and Das, S. K. (2008). Coverage and connectivity issues in wireless sensor networks: A survey, *Pervasive and Mobile Computing* **4**, 3, pp. 303–334.

[5] Heidemann, J., Stojanovic, M. and Zorzi, M. (2012). Underwater sensor networks: applications, advances and challenges, *Philosophical Transactions of the Royal Society A: Mathematical, Physical and Engineering Sciences* **370**, 1958, pp. 158–175.

[6] Kantor, G., Singh, S., Peterson, R. A., Rus, D., Das, A. K., Kumar, V., Pereira, G. A. S. and Spletzer, J. R. (2003). Distributed Search and Rescue with Robot and Sensor Teams, in *FSR*, pp. 529–538.

[7] Kumar, S., Lai, T.-H. and Arora, A. (2007). Barrier coverage with wireless sensors, *Wireless Netw.* **13**, 6, pp. 817–834.

[8] Lambrou, T. P. and Panayiotou, C. G. (2009). Collaborative area monitoring using wireless sensor networks with stationary and mobile nodes, *EURASIP Journal on Advances in Signal Processing* .

[9] Li, M., Cheng, W.-F., Liu, K., Liu, Y., Li, X.-Y. and Liao, X. (2011). Sweep coverage with mobile sensors, *IEEE Trans. Mob. Comput.* **10**, 11, pp. 1534–1545.

[10] Liu, B., Dousse, O., Nain, P. and Towsley, D. (2013). Dynamic coverage of mobile sensor networks, *IEEE Transactions Parallel Distributed Systems* **24**, 2, pp. 301–311.

[11] Lochmatter, T., Raemy, X. and Martinoli, A. (2007). Odor source localization with mobile robots, *Bulletin of the Swiss Society for Automatic Control* **46**, pp. 11–14.

[12] Mei, Y., Xian, C., Das, S., Hu, Y. C. and Lu, Y.-H. (2007). Sensor replacement using mobile robots, *Computer Communications* **30**, 13, pp. 2615–2626.

[13] Nikolaus, C. and Martinoli, A. (2009). Multirobot Inspection of Industrial Machinery From Distributed Coverage Algorithms to Experiments with Miniature Robotic Swarms, *IEEE Robotics & Automation Magazine* **16**, pp. 103–112.

[14] Palmer, D., James, G. and Corke, P. I. (2004). ElectricCOW: A Simulator for Mobile Sensors and Actuators Mounted on Herds of Cattle, in *Proc. 29th Annual IEEE Conference on Local Computer Networks (LCN'04)* (Tampa, FL, USA), pp. 556–557.

[15] Savkin, A. V., Javed, F. and Matveev, A. S. (2012). Optimal distributed blanket coverage self-deployment of mobile wireless sensor networks, *IEEE Comm. Lett.* **16**, 6, pp. 949–951.

[16] Sheu, J.-P., Hsieh, K.-Y. and Cheng, P.-W. (2008). Design and implementation of mobile robot for nodes replacement in wireless sensor networks, *Journal of Information Science and Engineering* .

[17] Tezcan, N. and Wang, W. (2007). Effective Coverage and Connectivity Preserving in Wireless Sensor Networks, in *Proc. IEEE Wireless Communications and Networking Conference (WCNC'07)* (Hong Kong, China), pp. 3388–3393.

[18] Wang, B., Lim, H. B. and Ma, D. (2009). A survey of movement strategies for improving network coverage in wireless sensor networks, *Computer Comm.* **35**, pp. 1427–1436.

[19] Xie, L., Shi, J., Hou, Y. T. and Sherali, H. D. (2012). Making sensor networks immortal: An energy-renewal approach with wireless power transfer, *IEEE/ACM Transactions on Networking* .

[20] Zhang, L., Tang, J. and Zhang, W. (2009). Strong Barrier Coverage with Directional Sensors, in *Proc. IEEE Global Communications Conference (Globecomm'09)* (Honolulu, Hawaii, USA), pp. 1–6.

[21] Zhu, C., Zheng, C., Shu, L. and Han, G. (2012). A survey on coverage and connectivity issues in wireless sensor networks, *Journal of Network and Computer Applications* , 35, pp. 619–632.

Chapter 5

Substitution Network: Controlled Mobility for Network Rescue

Isabelle Guérin Lassous[1,2], Tahiry Razafindralambo[1]

[1] *Inria*, [2] *Université Lyon 1, LIP (UMR ENS Lyon, CNRS, Inria, UCBL) France*

Abstract. *A substitution network is a rapidly deployable backup wireless solution to quickly react to network topology changes due to failures or to flash crowd effects on the base network. Unlike other ad hoc and mesh solutions, a substitution network does not attempt to provide new services to customers but rather to restore and maintain at least some of the services available before the failure. Furthermore, a substitution network is not deployed directly for customers but to help the base network to provide services to customers. Therefore, a substitution network is not, by definition, a stand-alone network. In this chapter, we describe the main challenges regarding the implementation of substitution networks that are related to some simple questions: 1) When to deploy the substitution network and when to stop the deployment? 2) Where to place the mobile substitution routers and where to place the network services that must be implemented? 3) How to implement the network mechanism and protocols in order to fit the base network and substitution network needs?*

5.1 Introduction and Context

We define a *substitution network* as a rapidly deployable backup wireless solution to quickly react to network issues due to failures or flash crowd effects on an existing network. Unlike other ad hoc and mesh solutions, a substitution network does not attempt to provide new services to customers but rather to restore and maintain at least some of the services available before the failure of the existing network. Furthermore, a substitution network is not deployed directly for customers but to help the existing network to provide services to customers. Therefore, a substitution network is not, by definition, a stand-alone network.

In this chapter, we focus on scenarios where nodes of an existing network, such as classical network routers, are disconnected or experience some issues, and connectivity or issue must be restored or solved between them by using wireless mobile routers whose mobility can be controlled. It is critical to design efficient algorithms to support these networked environments with requirements on the connectivity quality and, potentially, on Quality of Service (QoS) or Quality of Experience (QoE). Some of the main challenges include satisfying application-specific requirements under network link quality changes and without prior knowledge on the optimal wireless mobile router locations or the optimal network topology. This is a fairly complex node placement/deployment problem and constitutes the main focus of this chapter. Surprisingly, such a fundamental problem has received very few attention in the literature.

The goal of a substitution network is therefore to deploy a set of wireless mobile devices between classical network routers to restore, without prior knowledge of the optimal wireless mobile router locations, the connectivity, to improve the network performance, and, if possible, to satisfy applications' requirements, such as, for instance, delay, throughput or Mean Opinion Score. Substitution networks use *controlled mobility*, provided to wireless mobile routers, to improve network performance at all the layers of the protocol stack. Previous works have been done on deployment and placement of mobile devices (e.g., robots) for area coverage [3, 5, 21, 24]. But, in these works, the controlled mobility is exploited at the application layer to provide new services. In this chapter, controlled mobility is introduced at all layers of the protocol stack that can exploit this controlled mobility to enhance their own performance. We specifically aim at introducing mobility as a primitive of the protocol stack.

Based on this definition of a substitution network, we also assume that mobile devices should have *self-organization*, *self-optimization*, and *self-healing* capabilities in order to enable a flexible, scalable and resilient deployment. Hence, the main issue in this context is *when* and *where* to *deploy* or *re-deploy* the wireless mobile routers in order to provide the best performance for applications.

In this chapter, we first define the different underlying concepts of substitution networks and give the objectives of substitution networks. Then, we define the architectural and functional requirements for substitution networks. The scientific and technical challenges raised by substitution networks are then discussed. Finally, we describe some first solutions related to the scientific and technical challenges presented in the previous section. We also give, when possible, some recommendations to improve and/or to implement these solutions.

5.2 Objectives and Definitions

The deployment of temporary wireless mobile routers, called hereafter *mobile substitution routers,* can be useful to help an existing network, such for instance a backhaul network, that is virtually unreachable due to a surge of traffic, a failure of an equipment or a power outage, and that has no backup solution. In these cases, the deployment of a set of mobile substitution routers can help to restore the services to the subscribers. We call this set of deployed temporary mobile substitution routers a substitution network. Unlike conventional implementations of wireless LANs which rely on a fixed interconnection of access points (APs), substitution networks must be rapidly deployed in order to rescue existing networks which experience some troubles. Hereafter, the existing network is called *base network.*

A substitution network does not attempt to provide new services to customers but rather to restore and maintain at least some of the services available before the failure through the usage of mobile substitution routers. A substitution network is completely integrated inside the base network, even if it is not directly deployed with the base network. A substitution network takes also advantage of the controlled mobility properties of its entities to provide more flexibility and better performance during rescue operations.

Figure 5.1 shows an example of substitution network usage where a substitution network is deployed to support the failure of one equipment inside the base network. In Figure 5.1(a), there are two types of entity that compose the base network, the classic routers and the *bridge routers.* Unlike classic routers, bridge routers must have, at least, a wireless interface and are used as bridges between the base network and the substitution network. Therefore, bridge routers play an important role when the substitution network comes at play. It is worth noting that, in this figure, mobile substitution routers are already available but not used. Figure 5.1(a) shows the normal operation of the base network. In Figure 5.1(b), a failure occurs inside the base network. This failure can occur either on classic routers, bridge routers or links. In Figure 5.1(c), mobile substitution routers come at play to first restore the connectivity due to the failure inside the base network. Figure 5.1(d) depicts an example of the usage of controlled mobility for mobile substitution routers. This figure shows a movement of mobile substitution routers to increase the network performance. When the failure on the base network is solved, the substitution network through the usage of mobile substitution routers stops its operation and the base network can operate in "normal" mode such as in Figure 5.1(a).

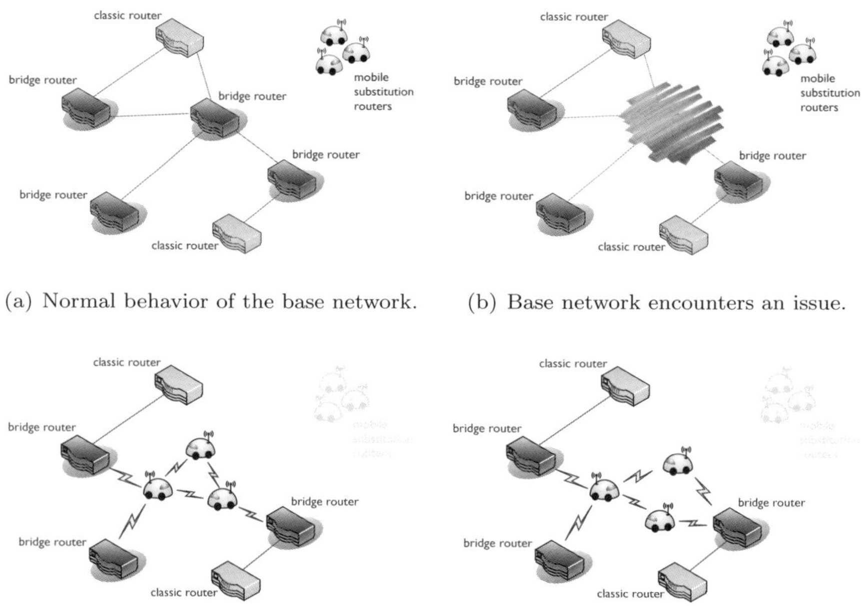

(a) Normal behavior of the base network. (b) Base network encounters an issue.

(c) Substitution network comes at play. (d) Substitution network operations.

Fig. 5.1 Example of substitution network usage.

5.3 Architectural and Functional Definitions

In this section, we define the architectural and functional requirements for substitution networks. This section highlights and gives the properties of the main hardware and software components for substitution networks to operate correctly. In this section, we differentiate the mandatory hardware and software components (denoted by (M) hereafter) from the optional ones (denoted by (O) hereafter) and rate the importance of such functionalities and/or hardwares. It is worth noting that, in this section, we try to be as general as possible and not to focus on specific hardwares, softwares, mechanisms or technologies.

5.3.1 *Hardware*

Since substitution networks involve the usage of specific hardware, it is important to first define the restrictions, limitations, requirements and assumptions related to these specific materials.

MOTION CAPABILITY (M): First, in order to limit the human intervention for the deployment of the substitution network, the mobile substitution routers must have motion capabilities. Wifibots[1] or micro-drones, like for instance AR.drone[2], are examples of devices that could be used as mobile substitution routers. Of course, mobile substitution routers must be able to operate on batteries with enough capacities for deployment of the substitution network to be useful. Note that, however, in this context, the depletion of a mobile substitution router's battery is not a dramatic event, as the mobile substitution router can be replaced by another mobile substitution router from the mobile substitution routers pool (if not empty) and fetched to be recharged in energy.

COMMUNICATION INTERFACE (M): Second, as the substitution network provides alternative paths to the base network, the mobile substitution routers and the bridge routers must be equipped with communication interfaces, and more specifically with wireless communication interfaces as the mobile substitution routers can move. The bridge routers correspond to some routers of the base network to whom a wireless interface (or more) has been added. As the substitution network has an unpredictable topology, that will depend on the location of the failure, it can be multihop. Therefore, the used wireless technology must be adapted to the multihop context. The communication interface should be able to send and receive data even if the full-duplex property is not mandatory. It is worth noting that the performance provided by the substitution network strongly depends on the performance of the communication interface used by the mobile substitution routers.

[1] `www.wifibot.com`

[2] `ardrone2.parrot.com`

Positioning Information (M): The position is an important feature that the mobile substitution routers should know. The positioning system enables the mobile substitution routers to retrieve their current position and to take appropriate decision on their movement. Without positioning information, the deployment of mobile substitution routers is much more difficult or even impossible. This positioning information can either be provided by internal network mechanisms, based, for instance, on signal strength, triangulation, or time difference of arrivals or by an external mechanism or hardware such as, for instance, GPS (Global Positioning System).

External Monitoring System (O): A monitoring system can be used to get an accurate view of the network and can be deployed with the substitution network to provide real time information about the status of the network. This system could provide information about the wireless spectrum usage and/or any other useful information regarding the radio frequencies used by or available for the substitution network. Such a system can be deployed as an external system such as the one described in [8] or appended to the mobile substitution routers. The monitoring system implies the usage of specific hardware such as, for instance, additional promiscuous wireless cards to provide the useful information.

5.3.2 *Software*

In this part, we describe all the software functions that are mandatory for the substitution network to operate. We also recommend some software functions, that are not mandatory for the substitution network, but that can improve the running and the performance of the substitution network (and in turn the traffic performance).

Protocol Stack Transition (M): A protocol stack transition method must be implemented between the substitution network and the base network. This transition method can be based on : 1) The co-existence of two protocol stacks on the bridge routers. In this case, the bridge routers "translate" the protocols of the base network to the protocols used by the substitution network at the entry of the substitution network and make the reverse "translation" from the substitution network to the base network. 2) The tunneling of base network traffic into the substitution network protocol stack. In any case, the protocol stack transition is a mandatory feature for the substitution network operation.

Monitoring Function (M): The deployment (and the redeployment) of the substitution network depends on the performance of the base network and the substitution network. It is thus important to provide an updated and accurate view of the performance and traffic inside both the base network and the substitution network. Therefore, a monitoring system of the base network and the substitution network has to be developed. To this end, a monitoring function can be integrated into bridge routers and mobile substitution routers. The monitoring function of the bridge router keeps track of traffic (and its associated performance) that crosses the bridge router from the base network to the substitution network (and vice versa). The monitoring function of the mobile substitution router keeps track of traffic (and its associated performance) that crosses the mobile substitution router, but it can also monitor surrounding traffic that can be sensed or captured on the wireless medium. The monitoring function can be implemented at different layers of the protocol stack, as for instance, the physical, MAC, network or even transport layer depending on the implementation of the Protocol Stack Transition mechanisms.

DEPLOYMENT ENGINE (M): According to the results of the monitoring function on the substitution network and the base network, the deployment engine of the bridge routers will trigger or stop the deployment of the substitution network. The deployment engine is used to control the deployment of the mobile substitution routers by adding or removing mobile substitution routers. This function is therefore very important as it is the core of the substitution network architecture. The deployment engine also gives self-configuration information to the mobile substitution routers.

MOBILITY ENGINE (M): The mobile substitution routers are equipped with a mobility engine. Used in an autonomous way by a mobile substitution router, it will enable a self-positioning of the mobile substitution router that can improve the substitution network performance. Only mobile substitution routers have mobility engine. It provides distributed and/or localized mechanisms and algorithms to enable the mobile substitution router to move in an autonomous way. The MOBILITY ENGINE can also be used to trigger a global behavior, such as for instance swarm behavior, inside the substitution network, only from local interactions between mobile substitution routers.

ROUTING/FORWARDING PROCESS (M): Since we do not make any specific assumption on the base network, the size of the substitution network and the number of mobile substitution routers are not known in advance. Therefore, the substitution network can be a multihop wireless network. In this case, an appropriate Routing/Forwarding process must be implemented within the substitution network. The routing process can either be the same as the one used in the base network or completely different.

Classical ad hoc networks routing protocols such as, for instance, OLSR [7], AODV [18] or geographic routing algorithms [25] can be used in the substitution network.

Distributed Medium Access Control (O): Since the substitution network is a distributed system by nature and whose the topology is likely to be multihop, some of the protocols used by the mobile substitution routers should be distributed or localized. The Medium Access Control (MAC) protocol, which is responsible of managing the channel access of wireless devices, is mainly concerned with this property. A distributed MAC protocol is recommended within the substitution network to avoid the management of channel access by a central entity on mobile devices.

Statistical and forecasting methods (O): The behavior of the substitution network strongly relies on the monitoring system which must provide enough data to generate an accurate view of the system. These data can also be used to predict the behavior of the networks (substitution network and base network) by using forecasting techniques. Forecasting can therefore be implemented to predict traffic growth or network performance collapses in order to avoid network issues.

Quality of Service (O): QoS is recommended to implement some QoS at the bridge routers and/or inside the substitution network. Some of the QoS services that could be implemented are:

- Traffic classification. Before entering into the substitution network, every single flow is assigned to a given class of priority that will determine how its packets will be handled within the substitution network.
- Admission Control. As the substitution network capacity is likely to be smaller than the base network capacity, admission control may be necessary. Admission control (implemented at bridge routers) aims at improving the overall quality of communications by preventing the network from congestion.
- Traffic control. The rate of the flows entering the substitution network can be controlled by using primarily traffic shaping, queue management and scheduling techniques.

5.4 Substitution Network: Scientific and Technical Challenges

The implementation of the requirements described in the previous section raises some scientific and technical challenges that can be summarized by the following questions: When? Where? How? In this section, we discuss these challenges and describe the fundamental issues linked to each of them by asking some fundamental questions. Behind each question, we try to answer to the two underlying questions, namely: What is the problem? And, why is it a problem?

5.4.1 *When?*

When to deploy/remove the substitution network? Substitution network is a rapidly deployable wireless network that is used as a backup support for an existing base network. This backup support is not installed directly with the base network but deployed on demand depending on spatial and temporal state of the base network. One one hand, since by definition, the substitution network has to be deployed rapidly, issues on the base network must be detected as quickly as possible and the deployment engine at the bridge routers must rapidly take the decision of deploying the mobile substitution routers on the field. On the other hand, the deployment engine at each bridge router must differentiate between short and long term issues. Indeed, since it takes some time to deploy the mobile substitution routers, if the issue on the base network is going to last less than the time needed to deploy the mobile substitution routers, the deployment engine at each bridge router must not deploy the mobile substitution routers on the field.

The fundamental problem here is to know whether the issue encountered by the base network is going to be short or long and to find the best trade-off between reactivity and deployment cost. Figure 5.2 plots different evolutions of the network health. We explicitly name it "health" to remain as general as possible. This health could be, for instance, the network performance and could be measured, for instance, by the overall throughput. It is important to notice here that Figure 5.2 only shows some examples and is not exhaustive. This figure only illustrates the fundamental issue raised by the "when" question. In Figure 5.2, h_1 is the maximum health of the network and h_0 is the minimum one.

Figure 5.2(a) is the simplest case. Indeed, in this figure, the substitution network deployment can be triggered at time t_1 when the base network encounters a sudden health issue. In Figure 5.2(b), the triggering of the substitution network deployment is less obvious. Indeed, substitution network deployment can be triggered at t_1, t_2 or at any moment between t_1 and t_2 which depends on the health's value h between h_0 and h_1. In Figure 5.2(c), the deployment of the substitution network can be divided into two cases and can be even more complicated. In this figure, if $t_2 - t_1 >> t_d$, where t_d is the time needed to deploy the substitution network, the deployment is easy to manage since the problem duration is long enough to make the deployment of the substitution network helpful. However, if $t_2 - t_1 \leq t_d$ the substitution network should not be deployed. In this case, the Deployment Engine should not trigger the deployment. Figure 5.2(d) is like the problem described in Figure 5.2(c) but in which the issue appears successively. In this case, if $t_{i+1} - t_i < t_d$ and $t_8 - t_1 > t_d$, the substitution network should be deployed at t_1 and should not be stopped before t_8. It is important to notice here that the deployment triggering and the deployment stop raise the same issues.

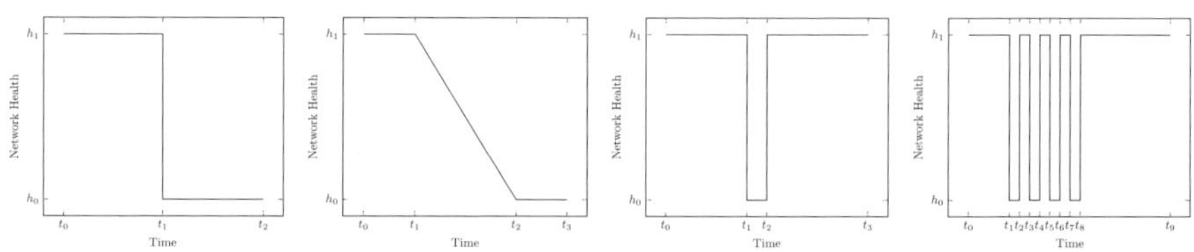

(a) (b) (c) (d)

Fig. 5.2 Example of the network health evolution.

In this section, we have highlighted a fundamental problem for substitution network from the deployment point of view. Each deployment, re-deployment or mobile substitution routers removal needs to be triggered carefully. In order to take the appropriate and most efficient "When" decision, the Deployment Engine must have 1) a 'real-time' view of the network's health and 2) prediction mechanisms to evaluate the duration of the encountered issues.

5.4.2 *Where?*

Where to place the mobile substitution routers when the deployment of the substitution network is triggered? When an issue is detected on the base network, the substitution network comes into play. The issue undergone by the base network is spatially and temporally located. Therefore, mobile substitution routers should be carefully placed and/or replaced to restore the health of the base network. As the mobile substitution routers will communicate by using a wireless technology, the position of each mobile substitution router is important since the radio frequency signal strength received at the destination decays in power law with the distance. Then, the received signal strength (RSS) can be directly linked to the signal over noise ratio (SNR) which is directly linked to the bit error rate (BER). The bit error rate can then be translated to packet error rate (PER) which, in turn, is a network performance indicator. The higher the packet error rate, the lower the successful data transmitted inside the network. Therefore, the wireless links performance has a clear impact on the network performance.

The fundamental problem here is then to know where to place the mobile substitution routers in order to increase the "quality" of each link. Here, the term quality is used to be as general as possible. Quality can be, for instance, the receive signal strength or the packet error rate or any measure that can be computed on a wireless link and that expresses the wireless link performance.

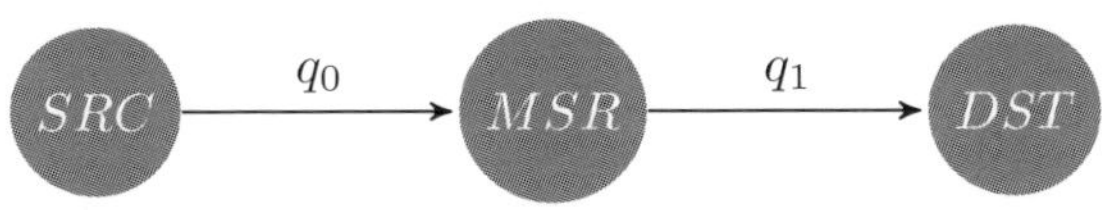

Fig. 5.3 Example of the network quality placement.

Figure 5.3 shows a simple example of a substitution network deployment. In this case, we can assume, without loss of generality, that SRC and DST are bridge routers. A mobile substitution router (denoted MSR) has been deployed between the two bridge routers to restore the network services. In this figure, q_i, $i \in \{0, 1\}$, represents the link quality. In this case, we assume that the mobile substitution router can evaluate q_i or that q_i can be provided by another entity of the network to the mobile substitution router. Let us assume that the maximum value of q_i is q_{max} (the best link quality) and its minimum value is q_{min} (the worst link quality). We assume also, as stated earlier, that the position of the mobile substitution router can modify the value of q_i.

In this context, the goal of the "Where" problem is to maximize the health of the network by modifying the position of the mobile substitution router. However, due to the wireless channel instability, the relationship between the position of the mobile substitution router and q_i is not a defined and known function. Moreover, q_i is a local measure, associated to a link, while the health of the network can be a global measure. It is thus not clear what the impact on the network health is when modifying q_i. From the mobile substitution router point of view, possible strategies are: 1) Equalize all the q_i or minimize their difference. 2) Maximize the minimum q_i to be Pareto optimal. 3) Depending on q_{max} and q_{min}, maximize the product of all q_i. 4) Minimize the difference between the $\max_i q_i$ and $\min_i q_i$. In all these cases, the important point is avoid being trapped within a local maxima/minima.

This "Where" problem can, of course, be generalized to any substitution network topology, more complex than the one depicted in Figure 5.3. The main difficulty of this problem is to find locally the best locations of the mobile substitution routers (based on local measurements of link quality) in order to achieve the best network health.

In the strategies described previously, we consider that only mobile substitution routers can take a movement decision by using the MOBILITY ENGINE. It is important to notice here that the computation of the mobile substitution routers' position can be made by a central entity (a bridge router of the substitution network for instance). However, in order to be efficient, this central entity must know the exact position of each mobile substitution router and the effect of the movement of each mobile substitution router on the links quality (or at least an approximation). Moreover, in order for the system to be responsive, the information about the links quality and the positions of the mobile substitution routers should be gathered by this central entity as fast as possible and the new mobile substitution routers' positions should also be transmitted from this central entity to the mobile substitution routers as fast as possible. This central entity must also not experience failure. And finally, even if the relationship between the mobile substitution router position and the link performance could be approximated, this approximation will only be valid for an unknown given period of time, in a specific (with unknown bound) geographic position and for a particular data traffic pattern. Therefore, the outcome of the mobile substitution routers' position computation may not be optimal. For these reasons, we think that a centralized approach for the "Where" decision is not a feasible approach.

Depending on the technology used, mobile substitution routers positions configuration may lead to some topological issues. For instance, if the IEEE 802.11 DCF (Distributed Coordination Function) method is used, the mobile substitution routers' configuration should avoid the hidden terminal effect or other configuration as described in [6]. With a code division multiple access (CDMA) method, topologies that could create near-far effect [19] should be avoided.

In this section, we have highlighted a fundamental problem for substitution network concerning the mobile substitution routers location. Each mobile substitution router must be positioned carefully. In order to take the appropriate and most efficient "Where" decision, the MOBILITY ENGINE must have: 1) an accurate evaluation of the link quality and 2) a mean to approximate/forecast the effect of its position on the link quality.

5.4.3 *How?*

This section shows some important challenges related to how things should be done within the context of substitution network. This section is not exhaustive, but we only try to highlight different challenges that we think as the most important.

How to get an accurate view of the network health or of the links quality? In order, for the DEPLOYMENT ENGINE and the MOBILITY ENGINE, to take the appropriate decisions at the right time, the network health and the links quality should be made available in real time to the bridge routers and the mobile substitution routers. This knowledge is very important for the substitution network since the main decisions are based on these two measures.

How to coordinate the mobile substitution routers movements inside the substitution network? Since the MOBILITY ENGINE is responsible for the movement of each mobile substitution router, it is also responsible of the coordination of the mobile substitution routers' movements. Indeed, let us assume that the interactions between mobile substitution routers can be represented by a graph where the vertices are the mobile substitution routers and the edges represent possible wireless communications between mobile substitution routers. This graph should have at least the connectivity property. That is, if two mobile substitution routers (neighbors in the graph) decide to move apart from each other, the communication link between them can be broken due to the distance effect on the received signal strength. But, in this case, the mobile substitution routers should coordinate in order to maintain the graph connectivity during their movements.

How to forecast the link quality depending on the mobile substitution router's movement? The movement of each mobile substitution router has a clear impact on the links quality in its surroundings. It is thus important to know what the impact of each movement of a mobile substitution router is in order to anticipate the movement decision that should be made, but also to avoid the mobile substitution router to be trapped in a local minima/maxima. Therefore, it is important for each mobile substitution router to know and to forecast the effect of its movements on the link quality depending on the context (including traffic, its position, the position of neighbor mobile substitution routers, etc.).

How to make the substitution network operational with good performance? As explained previously, the substitution network is likely to be a multihop wireless network, with specific features as the traffic will enter into the substitution network at one or several bridge routers and will leave the substitution network at one or several bridge routers. The used routing protocol should be adapted and efficient for these kinds of topologies. Moreover, the substitution network is likely to have a smaller capacity than the base network that is in trouble, as the routers inside the substitution network communicate via a wireless technology. Therefore, a smart management of the traffic will be important in the substitution network in order to ensure good performance. All the traffic transmitted in the base network may not be transmitted in the substitution network or may not be transmitted with the same traffic profile. Mechanisms such as admission control based on flow priority, traffic shaping, multi-path routing and/or load balancing can be considered. These mechanisms will help the substitution network to operate in a "safe-mode" and must avoid possible collapses.

How to forecast network traffic/network health on the substitution network and the base network? The network traffic or the network health forecasting is as important as the knowledge of an accurate view of the current network traffic or health. Indeed, the proposed system must be responsive in order to minimize the impact of the issues that could appear. The forecasting techniques can, for instance, minimize the effect of a traffic surge or network health collapse. Some forecasting techniques on the network traffic or network health will enable a switch from a completely reactive substitution network paradigm to a proactive substitution network paradigm and therefore avoid denial of services.

5.5 First Implementations and Recommendations

In this section, we review and highlight some results from the literature that are related to the scientific and technical challenges described in the previous sections. We mainly focus on the methods and techniques described in the literature that provide some (even partial) solutions to one or more of the issues raised in the previous sections. We also point out the weaknesses and strengths of the proposed solutions and, when possible, give some recommendations to improve and/or to implement these solutions (on specific hardwares if appropriate).

Feasibility and use cases: In [22, 26], the authors describe possible use cases of substitution network and its associated mechanisms. These papers describe possible hardwares and uses. The authors, for instance, describe specific use cases, ranging from animal tracking in a farm to disaster recovery at a city scale. The authors of [20] give a detailed recommendation of a QoS architecture implementation that should be implemented in substitution network. They describe the QoS mechanisms and provide the way to implement them within the substitution network context. In [11], the authors review some testbed implementations that can be used as an example of substitution network implementation in a "in vitro" environment.

Hardware: In [23], the authors describe and give the characteristics of commercial flying drones that could be used as mobile substitution routers. The proposed hardware consists of off-the-shelf technologies with, for example, IEEE 802.11b as a communication interface, quadrotors as a mean of mobility and GPS as a mean of localization. In [21], the authors propose another alternative by using moving robots. The robots use only commercial hardware such as 802.11b communication interface, classic embedded PC running a classic Linux distribution, etc. The authors of [21] have also experimented some of the substitution network mechanisms using the described robots.

Accurate view of the network: In [4, 8, 16], the authors propose monitoring architectures that can provide useful and accurate information about the network state. In [4, 16], the proposed monitoring architecture mainly focuses on a cross-layer (from the PHY layer to the network layer) estimation of the wireless channel quality and use. In [8], the proposed architecture focuses on the MAC layer and tries to capture the network activity from a frame point of view. The proposed architecture is not yet ready to be fully integrated in a substitution network. Indeed, the capture method is already proved to operate correctly but the online traces and data analysis is still under development.

Mobility engine: The papers [10, 21] give some simulation and experimental results regarding the implementation of the MOBILITY ENGINE. In the proposed experiments, the authors have succeeded in implementing a distributed swarm behavior on mobile robots. More specifically, the authors propose an algorithm that can autonomously deploy mobile substitution routers between two bridge routers. The proposed algorithm is also robust against robots failure and automatically restores connectivity within the substitution network in case of a robot failure. It is important to notice here that the papers [21] and [10] do not focus on the improvement of link quality between mobile substitution routers. Indeed, these works mainly focus on the "initial" deployment of mobile substitution routers between two or more bridge routers and robustness of this deployment against mobile substitution routers failures, as illustrated by Figure 5.1(c). Figure 5.4 (extracted from [10]) shows an example of a deployment experimentation of three mobile substitution routers between two bridge routers. In this experiment, the communication range of mobile substitution routers and bridge routers has been virtually reduced. The experimentation shows that the proposed algorithm implements a simple obstacle avoidance and that, during the whole deployment procedure, the connectivity between the starting bridge router and the mobile substitution routers is always kept.

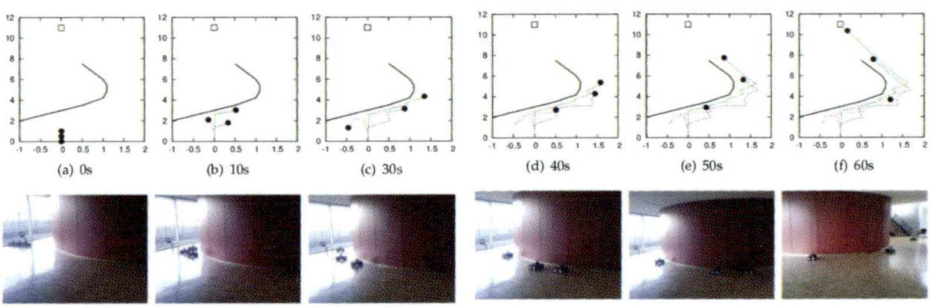

Fig. 5.4 Deployment of three mobile substitution routers between two bridge routers. The connectivity is kept all along the deployment procedure and an obstacle avoidance mechanism is implemented.

Mobile substitution routers' replacement: The papers [12–14] propose simple placement mechanisms of mobile substitution routers. The proposed algorithms are based on different link quality measures such as delay, RSSI, etc. For each link quality measure, the associated algorithm, operating on each mobile substitution router, seeks to equalize the measure between all the surrounding links of the mobile substitution router. The evaluation of the different algorithms is carried out by simulation and relies on video transmission quality assessment. The obtained results show that simple measures, as the ones considered, can increase the end-to-end performance of the substitution network and that the performance of the substitution network increases all along the deployment procedure. However, the results presented in [14], [12] and [13] show that no single metric, among the ones studied, outperforms the others and provides the optimal mobile substitution routers replacement. In Figure 5.5 (from [14]), we show a simple example of the evolution of a mobile substitution router's position based on the delay. In this experimentation, a mobile robot is used for mobile substitution router. The experiment is done with two bridge routers and one mobile substitution router that acts as a relay (such as in Figure 5.3). The link quality measure is the delay and the mobile substitution router tries to equalize the delay of the two links. The x-axis of the figure is the time and the y-axis in the position (in meters). The mobile substitution router starts at position 0 close to one of the bridge router. This figure shows how the mobile substitution router constantly tries to optimize its position (the red-curve corresponds to a step-wire positioning, while the grey curve corresponds to a continuous positioning). In [17], another placement mechanism, based on available energy on the mobile substitution routers, is used. The three works could be merged in order to have an energy efficient and network efficient mobile substitution routers placement. The papers [14], [12], [13] and [17] mainly focus on the replacement of mobile substitution routers, as illustrated by Figure 5.1(d).

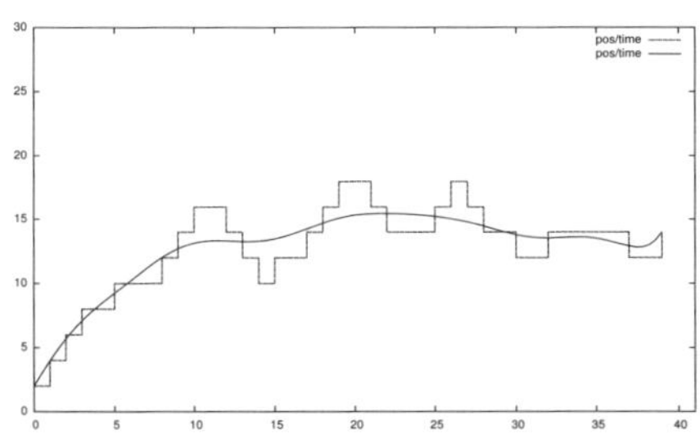

Fig. 5.5 Mobile substitution router position in a configuration with 2 bridge routers and 1 mobile substitution router. Replacement decision is taken based on delays.

Performance and Forecasting: In [1], the authors analyze the performance of the simplest possible multihop topology for a substitution network subject to traffic transmitted in both directions. This simplest topology is clearly a multihop chain. If the multihop chain has been extensively studied, especially when IEEE 802.11 technology in DCF mode is used as MAC protocol, most of these studies consider the case of a one-way traffic transmitted from one endpoint (i.e., bridge routers) of the chain to the other endpoint chain. But in a substitution network, the transported traffic is expected to be of roughly equal size in both directions. The evaluation is carried out by simulation under different traffic intensities and chain lengths. One result, given in Figure 5.6 (from [1]), shows that the overall throughput, achieved by the network (y-axis of Figure 5.6), reaches a peak at a given workload λ (x-axis of Figure 5.6) and then, for larger values of workload, decreases towards an asymptote which value can be drastically lower than the peak value. This performance collapse shows the need to forecast the evolution of the traffic and the network performance, but also the need for a suitable admission control mechanisms.

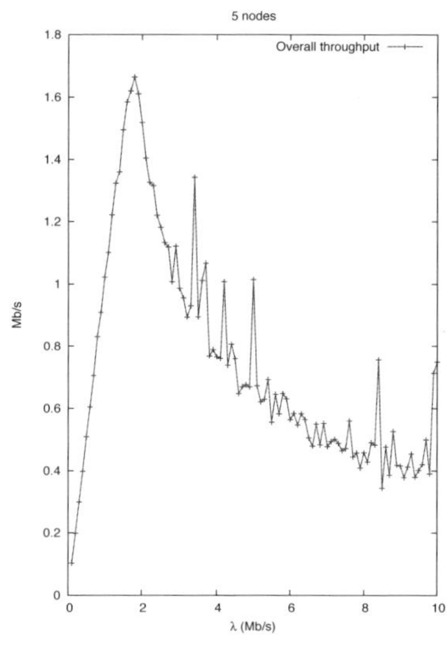

Fig. 5.6 Overall throughput on a 4-hop chain with UDP two-way traffic and with wireless links capacity of 11 Mb/s.

Protocols for substitution network: As mentioned previously, the network performance can decrease drastically if no care is taken. It is thus important to deploy efficient protocols for substitution networks. Multihop wireless networks face several issues, especially when IEEE 802.11 technology in DCF mode is used as MAC protocol [6]. These issues could be avoided if the topologies that raise these issues were discovered and the suitable protocols were used. In [15], the authors study the idle periods durations distribution in 802.11-based multihop wireless networks in order to determine whether the idle periods duration is a discriminating parameter that enables the identification of the topologies that raise issues. If the answer to this question is a partial answer in [15], the authors show how the idle periods durations distribution can be used to improve the performance of the MAC layer in mutlihop wireless networks. In [9], the authors study and improve the recovery properties of single and multipath routing strategies when facing network failure situations, specifically when two nodes (e.g., bridge routers) are connected via several paths, which could be the case in substitution networks. In these networks, mobile substitution routers may face failure simply due to a bad placement. It is thus important to have fast recovery strategies. In [9], the authors propose three new different recovery schemes and accordingly extend original routing solutions, like OLSR and MP-OLSR (Multi-Path OLSR), in order to decrease the expected latency and improve the protocols performance. Simulation results show a steep decrease of the latency when using these new schemes in dual chain topologies. Another possibility to improve the network performance is to add several wireless communication interfaces to the routers of the substitution network. However, to make the multi-radio feature effective, it is important to design solutions that manage efficiently the network resources. In [2], the authors propose a bandwidth reservation mechanism and an admission control solution for multi-radio multi-channel multi-hop networks based on the estimation of inter-flow and intra-flow interferences. The simulation results show the interest of such approaches.

5.6 Conclusion

In this chapter, we introduce the concept of substitution network. A substitution network is a rapidly deployable backup wireless solution to quickly react to issues of existing networks, called base networks. The substitution network is not directly deployed for customers or end users but only to help a base network that experiences issues. The substitution network is, therefore, not a stand-alone network and can rely on mechanisms implemented inside the base network or at the bridge routers which serve as bridges between the base network and the substitution network.

The main challenges regarding the implementation of substitution networks are related to some simple questions: 1) "When" to deploy the substitution network and when to stop the deployment? 2) "Where" to place the mobile substitution routers and where to place the network services that must be implemented? 3) "How" to implement the network mechanism and protocols in order to fit the base network and substitution network needs?

This chapter covers the main scientific, technical and implementation challenges raised by the usage of substitution network. It also describes possible use cases and provides recommendations for the mechanisms' implementation and for the hardwares that could be used. This chapter also reviews the literature on existing solutions that could be of interest for substitution networks and highlights their weaknesses and strengths.

References

[1] Abreu, T., Nguyen, N., Begin, T., Guérin Lassous, I. and Baynat, B. (2012). Substitution networks: Performance collapse due to overhead in communication times, in *Proceedings of the 4th International Conference on Ad Hoc Networks (AdHocNets)* (Paris, France).

[2] Ashraf, U., Abdellatif, S. and Juanole, G. (2012). Interference-aware bandwidth reservation in multi-radio multi-channel mesh networks, *Computer Communications* **35**, 17, pp. 2138 – 2149.

[3] Batalin, M. A. and Sukhatme, G. S. (2004). Coverage, exploration and deployment by a mobile robot and communication network, *Telecommunication Systems Journal, Special Issue on Wireless Sensor Networks* **26**, pp. 2–4.

[4] Carvin, D., Owezarski, P. and Berthou, P. (2012). Managing The Upcoming Ubiquitous Computing, in *Proceedings of the 1st Conference on Network and Service Management (CNSM)* (Las Vegas, USA).

[5] Chatzigiannakis, I., Kinalis, A. and Nikoletseas, S. (2006). Sink mobility protocols for data collection in wireless sensor networks, in *Proceedings of the 4th ACM international workshop on Mobility management and wireless access (MobiWac)* (Terromolinos, Spain), pp. 52–59.

[6] Chaudet, C., Dhoutaut, D. and Guérin Lassous, I. (2005). Performance issues with ieee 802.11 in ad hoc networking, *Communications Magazine, IEEE* **43**, 7, pp. 110 – 116.

[7] Clausen, T. and Jacquet, P. (2003). Optimized link state routing protocol (olsr), in *RFC Editor*, p. RFC 3626.

[8] Claveirole, T. and Dias de Amorim, M. (2010). Wipal: efficient offline merging of ieee 802.11 traces, *SIGMOBILE Mobile Computing Communication Review* **13**, 4, pp. 39–46.

[9] Doghri, I., Reynaud, L. and Guérin Lassous, I. (2011). On the recovery performance of single- and multipath olsr in wireless multi-hop networks, in *Proceedings of the 3rd International Conference on Ad Hoc Networks (AdHocNets)* (Paris, France).

[10] Erdelj, M., Razafindralambo, T. and Simplot-Ryl, D. (2013). Covering points of interest with mobile sensors, *IEEE Transactions on Parallel and Distributed Systems* **24**, 1, pp. 32 –43.

[11] Gluhak, A., Krco, S., Nati, M., Pfisterer, D., Mitton, N. and Razafindralambo, T. (2011). A survey on facilities for experimental internet of things research, *IEEE Communications Magazine* **49**, 11, pp. 58 –67.

[12] Miranda, K., Natalizio, E. and Razafindralambo, T. (2011). On the impact of router's mobility on substitution networks, in *Proceedings of the 12th ACM International Symposium on Mobile Ad Hoc Networking and Computing (Poster Abstract) (MobiHoc)* (Paris, France).

[13] Miranda, K., Natalizio, E. and Razafindralambo, T. (2012a). Adaptive deployment scheme for mobile relays in substitution networks, *International Journal of Distributed Sensor Networks* **2012**.

[14] Miranda, K., Natalizio, E., Razafindralambo, T. and Molinaro, A. (2012b). Adaptive router deployment for multimedia services in mobile pervasive environments, in *Pervasive Computing and Communications Workshops (PERCOM Workshops), 2012 IEEE International Conference on (Percom)* (Lugano, Switzerland), pp. 471 –474.

[15] Nguyen, N. V., Guérin-Lassous, I., Moraru, V. and Razafindralambo, T. (2012). Characterisation and application of idle period durations in ieee 802.11 dcf-based multihop wireless networks, in *Proceedings of the 15th ACM international conference on Modeling, analysis and simulation of wireless and mobile systems (MSWiM)* (Paphos, Cyprus), pp. 277–284.

[16] Owezarski, P., Hasan, R. G., Kremer, G. and Berthou, P. (2011). First step in cross-layers measurement in wireless networks: how to adapt to resource constraints for optimizing end-to-end services? in *Proceedings of the 9th IFIP TC 6 international conference on Wired/wireless internet communications (WWIC)* (Barcelona, Spain), pp. 150–161.

[17] Pace, P., Loscri, V., Natalizio, E. and Razafindralambo, T. (2011). Nodes placement for reducing energy consumption in multimedia transmissions, in *Proceedings of the IEEE 22nd International Symposium on Personal Indoor and Mobile Radio Communications (PIMRC)*, pp. 909 –914.

[18] Perkins, C., Belding-Royer, E. and Das, S. (2003). Ad hoc on-demand distance vector (aodv) routing, in *RFC Editor*, p. RFC 3561.

[19] Rappaport, T. (2001). *Wireless Communications: Principles and Practice*, 2nd edn. (Prentice Hall PTR, NJ, USA).

[20] Razafindralambo, T., Begin, T., Dias De Amorim, M., Guérin Lassous, I., Mitton, N. and Simplot-Ryl, D. (2011). Promoting Quality of Service in Substitution Networks with Controlled Mobility, in *Proceedings of the 10th International Conference on Ad Hoc Networks and Wireless (AdHocNow)* (Padderborn, Germany), pp. 248–261.

[21] Razafindralambo, T. and Simplot-Ryl, D. (2011). Connectivity preservation and coverage schemes for wireless sensor networks, *IEEE Transactions on Automatic Control* **56**, 10, pp. 2418–2428.

[22] Reynaud, L. and Rasheed, T. (2012). Deployable aerial communication networks: challenges for futuristic applications, in *Proceedings of the 9th ACM symposium on Performance evaluation of wireless ad hoc, sensor, and ubiquitous networks (PE-WASUN)* (Paphos, Cyprus), pp. 9–16.

[23] Reynaud, L., Zaimi, S. and Gourhant, Y. (2012). Competitive assessments for hap delivery of mobile services in emerging countries, in *Proceedings of*

The 3rd International Workshop on Business Models for Mobile Platforms (BMMP) (Berlin, Germany).

[24] Srinivasan, W. W. V. and Chua, K.-C. (2007). Trade-offs between mobility and density for coverage in wireless sensor networks, in *Proceedings of the 13th annual ACM international conference on Mobile computing and networking (MobiCom)* (Montreal, Canada), pp. 39–50.

[25] Stojmenovic, I. (2002). Position-based routing in ad hoc networks, *IEEE Communications Magazine* **40**, 7, pp. 128 –134.

[26] Venmani, D., Gourhant, Y., Reynaud, L., Chemouil, P. and Zeghlache, D. (2012). Substitution networks based on software defined networking, in *Proceedings of the 4th EAI International Conference on Ad Hoc Networks (AdHocNets)* (Paris, France).

Chapter 6

Energy Restoration in Mobile Sensor Networks

Nicola Santoro and Elio Velazquez

School of Computer Science, Carleton University, Ottawa, Canada

Abstract. *Restoration of sensors' energy is crucial to ensure continuous operations of a sensor network. In static sensor networks, energy restoration requires the use of special mobile entities (e.g., robots, actuators). In mobile sensor networks, the sensors themselves can solve the problem by moving to recharge facilities deployed throughout the sensing area. An efficient energy restoration strategy is one in which the losses are limited in scope and time; that is, with the minimum number of sensor losses, it allows the system to reach a state of equilibrium guaranteeing continuous operation of the network without any further sensor losses. In this Chapter, we describe and examine the basic strategies for energy restoration, passive and active, and analyze their performance.*

6.1 Introduction

6.1.1 *Energy management in sensor networks*

Regardless of the specific application being addressed, the ultimate goal of any sensor network is to achieve accurate sensing and maximize lifetime while maintaining an acceptable level of coverage. Since every sensor operation consumes energy, in any wireless sensor network deployment, the sensors' batteries will eventually deplete and loss of coverage will occur.

The most simplistic approach to cope with the eventual loss of coverage has been to deploy a very large number of sensors to compensate for the loss of the depleted ones. In this approach, the spare sensors must detect when a sensing hole is created by the battery depletion of a sensor and take its place. A large number of studies have indeed been dedicated to these problems (e.g., see [1, 2]). However, for obvious environmental reasons, these kinds of solutions are not really sustainable; furthermore, regardless of the number of spares, a loss of coverage over time is inevitable since there are no provisions to recharge or replace sensors in the long run.

The study of *energy management* has mostly focused on balancing the energy levels among all sensors (e.g., [3, 4]), rather than on energy restoration. For example, since sensors closer to the base station have to route/aggregate data flowing from remote parts of the network towards the base station, they tend to deplete their batterie much faster than other sensors; to overcome this disparity, a *mobile* base station could be used so to extend network operating life [3].

The crucial and constant concern, in every aspect of a sensor network design, on the minimization of energy consumption only ensures to delay the time of loss of coverage. But, in absence of provisions to recharge sensors, the network will eventually fail. The focus of this chapter is rather on *energy restoration* and providing *continuous operation* of the network.

6.1.2 *Energy restoration in static sensor networks*

To achieve *energy restoration* in static sensor networks, one approach is to have the sensors extracting energy from the environment (e.g., [5, 6]). The alternative approach to energy restoration in static sensor networks is to employ special entities to recharge the depleted sensors; these special entities could be particularly equipped sensors or dedicated robots; this approach is advocated also for network repairs where the special maintenance entities replace depleted sensors (e.g., [7–10]). For example, in [11], static sensors are recharged by chargers or actuators carrying solar panels.

In general, these energy restoration strategies can be categorized into two groups with some degree of overlap: *cluster-based* approaches (e.g., [12–15]) and *mobility-based* approaches (e.g., [3, 4, 7, 16, 17]). Indeed, most of these solutions rely on some kind of clustering or partitioning of the network with the special entities as cluster heads; they operate by either creating a fixed partition of the field or by constructing and maintaining dynamic clustering structures which depend on the current position of the cluster heads (e.g., [8, 10]). In each cluster, the special entity must obtain information about the energy distribution in the cluster by collecting additional information embedded in each communication with the static sensors (e.g., maximum energy, remaining energy, location, etc.) to decide when and where to intervene.

Other examples of these approaches can be found in research papers in robotics [11, 18, 19]. In all cases, a charger robot is responsible for delivering energy to a swarm of sensors. The recharging strategy is completely reactive (i.e., sensors are only recharged when they become out of service and cannot move). In the scenario described in [18], the charger robot is equipped with several docking ports. However, the charger robot can travel to recharge a robot in need only if none of the docking ports are occupied, assuming that several depleted robots need to be close by in order to be recharged simultaneously. The solution presented in [19], where a team of mini-robots (our sensors) are deployed along with more powerful docking station robots, is based on the creation of clusters of mini-robots.

6.1.3 *Energy restoration in mobile sensor networks*

All the above mentioned approaches for energy restoration are tailored for traditional sensor networks, that is where the sensors are *static*. The situation is drastically different in the case of *mobile* sensor networks, that is where the sensors are endowed with motor capabilities. Mobility clearly enhances the system allowing to perform tasks impossible in static sensor networks such as *self-deployment* [20–22], *aerial and maritime unmanned applications* [23–25].

In the case of mobile sensor networks, the sensors themselves can solve the energy restoration problem by moving towards recharge facilities deployed throughout the sensing area; in other words, the responsibility for maintaining the overall health of the network can be shifted to the sensor side, whereas the service facilities can play a passive role.

To ensure a *continuous* operation of a mobile sensor network, the mobile sensors are responsible for managing their own energy levels and for coming up with strategies to extend their operating life beyond one battery charge. The standard method to decide when to recharge has been based on fixed thresholds [17]. In this case, the service stations take a more passive role and the sensors should be able to compute their remaining operational time and coordinate the use of the service stations [16]. Furthermore, for instances where the sensors have to visit a predefined number of points of interests, [17] describes threshold vs. non-threshold-based solutions where robots decide to visit the service stations depending on their proximity and the nature (locations) of the points of interests.

The problem of achieving continuous operation by refuelling or recharging mobile entities has been the focus of attention in recent research papers in robotics. In particular, a general version of this problem is the *Frugal Feeding Problem* (FFP), so called for its analogy with occurrences in the animal kingdom [26, 27]. The FFP consists to find energy-efficient routes for a mobile service entity, called "tanker", to rendezvous with every member of a team of mobile robots. The FFP has several variants depending on where the "feeding" or refuelling of the robots takes place: at each robot's location, at a predefined location (e.g., at the tanker's location) or anywhere. Regardless of which variant is chosen, the problem lies in ensuring that the robots reach the rendezvous location without "dying" by energy starvation during the process. The context of mobile sensors and static recharge facilities deployed throughout the sensing area corresponds to the "tanker absorbed" version of FFP: the "rendezvous" between the recharging facility (the tanker) and the mobile sensors (the robots) takes place at the location of the recharging facility.

The problem of where to place a service facility is examined in [28] for mobile robots. In this case, a team of mobile robots have the specific task of transporting certain items from a pick-up to a drop-off location. To be able to work for a prolonged period of time, the robots should interrupt their work and visit the recharge station periodically (i.e., tanker-absorbed FFP). Their solution is to place the charger station close enough to the path followed by the robots but without causing interference to the robots' movements.

In all the aforementioned scenarios there are some necessary conditions for the sensors to be able to recharge themselves [18]. First of all, the sensors must be able to monitor their energy levels and detect when it is time to recharge. Second, they must be able to locate and move towards a charging station. Finally, there must be a mechanism for the energy transfer either by docking or plugging into the charging station or via wireless recharging at short distances (e.g., [16, 29–31]).

The *perfect* energy restoration strategy should be able to guarantee a continuous operation of the network without any losses; however in reality some sensor losses will occur. A *successful* energy restoration strategy is one in which the losses are limited in scope and time; that is, the strategy allows the network to reach a state of equilibrium, where no further sensor losses will occur (thus guaranteeing a continuous operation of the network), with the minimum number of sensor losses. The basic strategies for energy restoration in mobile sensor networks using static recharge facilities deployed throughout the sensing area have been introduced in [32, 33]. In the rest of this Chapter we will describe and analyze these strategies.

6.2 Basic Terminology and Assumptions

The system is composed of a set $S = \{s_1, ..., s_N\}$ of N mobile sensors distributed in an area of unspecified shape and a set $F = \{f_1, ..., f_K\}$ of K static recharge facilities distributed throughout the area. Each facility is equipped with a fixed number of recharging ports or sockets; this represents the maximum number of simultaneous sensors at the facility.

The placement of the service facilities can be achieved using any of the clustering algorithms shown in [34–38]. Once the clustering creation is finalized, there will be exactly one recharging station for each sensor in S. The sensors will know the location of their recharging facility, but the facilities are not required to know the number of sensors that will use their resources. With a clustering structure already in place, we can focus on the interactions within a particular cluster. Therefore, without loss of generality, our strategies will be presented in the context of one facility and the subset of mobile sensors assigned to its cluster.

The location of the sensors is assumed to be final in terms of their sensing assignment; in other words, from the point of view of the application (i.e., functional requirements), the sensors are already placed in the desired positions, e.g., though an initial self-deployment phase [20, 22, 39]. This means that if a sensor decides to move (e.g., to go to a service station to recharge its battery), it might create a sensing hole.

It is assumed that sensors can determine their own positions by using some localization method (e.g., GPS). Sensors can communicate with other sensors within their transmission range R and they all move at the same speed. The distance to the assigned facility should be within the sensors' mobility range to guarantee a successful round-trip to the station with one battery charge.

All communications are asynchronous; there is no global clock or centralized entity to coordinate communications or actions. The communication environment is assumed to be contention and error free (i.e., no need to retransmit data) and there is no interference produced by receiving simultaneous radio transmissions (i.e., ideal MAC layer).

To receive service from a recharge station $f \in F$, a sensor $s \in S$ must request the station f for a free socket. Once the request is granted, the sensor can then move to the station and recharge at the assigned socket.

There are two basic types of strategies for the energy restoration problem in our context: passive and proactive.

In a *passive* strategy, each sensor will monitor its energy level at periodic intervals and after any operation (e.g., send, receive, etc.). When its battery reaches a critical level, it goes to its recharging station.

In a *proactive* strategy, a sensor may decide to move before its battery reaches a critical level; the general idea is that sensors will try to get closer to their service stations in order to capture the so called "front seats" for when their time comes to make a trip to recharge their batteries.

6.3 Passive Approach to Energy Restoration

In a passive strategy, the sensors operate in two basic states: BATTERY_OK and BATTERY_LOW. Once the battery level falls below a predefined threshold (state BATTERY_LOW), which is not necessarily the same for all the sensors and depends on their distance to the station, the sensor must recharge its battery.

If the recharge station is within the sensor's transmission range, the sensor can send a recharge request directly and, once the request is granted, it moves to the assigned dock of the recharging station. If the recharging station is outside the sensor's transmission range, the request can be sent using some routing mechanism to forward the recharge request message to the service station. Alternatively, the sensor could start its journey towards the recharge station and once it gets there (or at least within range) it requests an available socket.

Regardless of the mechanism chosen, the sensor-facility interactions are implemented based on the service station communication pattern shown in Figure 6.1. For simplicity, the pattern shows the case of a service station with only one recharge socket. The recharging process is initiated with a RECHARGE_REQUEST sent by a low battery sensor. The service station will keep a queue of received requests and a ranking based on the sensors' energy levels. When a socket becomes available, the service station sends a RECHARGE_ACCEPT to the smallest ranked sensor (i.e., lowest energy sensor). Every time a sensor recharge is completed, the sensor sends a RECHARGE_DONE message to the service station and travels back to its assigned position in the network. This process is repeated continuously.

The effectiveness of this method depends on several factors such as: number of sensors in the cluster, distance to the station, number of recharg-

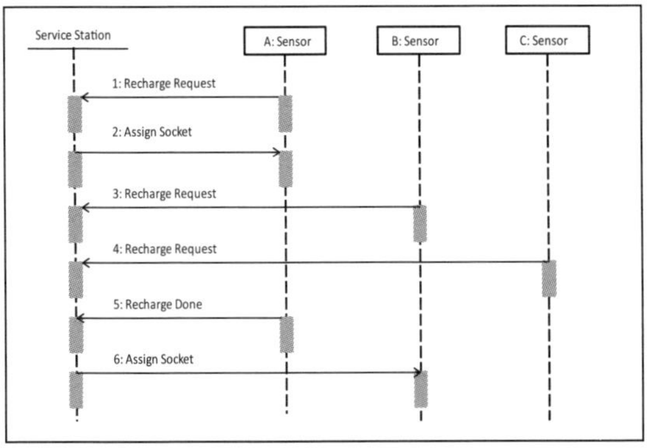

Fig. 6.1 Communication pattern for a mutex service station.

ing sockets, etc. Since our ultimate goal is to achieve a point of equilibrium with minimal or no sensor losses, a new question arises: will this approach work, and if it does, at what cost? The experimental analysis section provides some of these answers.

It is important to point out that, when sensors travel to the stations, they create temporary coverage holes. If temporary loss of coverage is an issue of paramount importance for the network, there are solutions to overcome this limitation. For instance, the service stations could be equipped with spare sensors. The number of spare sensors should be equal to the number of recharging sockets and every time a sensor is accepted (i.e., a socket becomes available), a spare is dispatched to the sensor's location to take its place. The low battery sensor is now free to travel to the base station and will eventually become a spare after its battery has been recharged.

6.4 Proactive Approach to Energy Restoration

In this section we examine the case when the mobile sensors decide to act before their batteries reach a critical level and a trip to the recharging station is needed. The general idea is that sensors will try to get closer to their service stations in order to capture the so called "front seats" (i.e., sensor locations within one-hop distance to the station). However, the number of front seats is limited and, since the sensors have functional responsibilities in their assigned locations and any movement can create a coverage hole, changing location cannot be a unilateral decision.

6.4.1 *Position based movements*

To minimize coverage holes (due to movements or to total energy depletion), a sensor with low energy will perform a gradual approach towards the recharge station, and it will do so by swapping position with higher energy sensors closer to the recharge station. The operating life of a sensor is now divided in three stages depending on its battery status: 1) BATTERY_OK or normal operation, 2) BATTERY_LOW or energy-aware operation, and 3) BATTERY_CRITICAL or recharge-required operation.

A sensor in a BATTERY_OK state will perform its regular sensing functions as well as accept any swapping proposal from other sensors with less energy. When the battery level falls below a fixed threshold, the sensor switches its state to a more active BATTERY_LOW state. In this state, the sensor will start its migration towards the recharge station, proposing a position swap to sensors with higher energy levels. Finally, a sensor in the BATTERY_CRITICAL state will contact the recharge station and wait until a socket or docking port has been secured; then, it will travel to the station and recharge (see Figure 6.2).

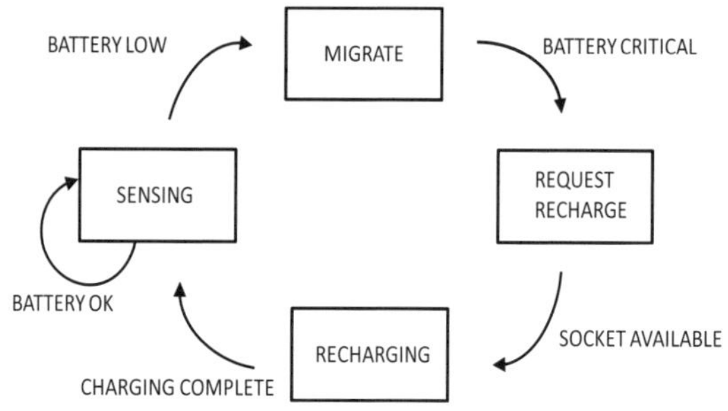

Fig. 6.2 A sensor's life cycle.

In this life cycle, it is the migration behavior that is of interest. The objective of the sensor during migration is to reach the recharge facility in an effective, timely manner, while relying solely on local information.

This can be done by allowing the sensor to explore energy-aware routes leading to the recharge facility. We propose to make use of position-based routing strategies (e.g., see [40, 41]). However, instead of sending a packet that needs to be routed until it reaches the intended target, the sensors have to "route themselves" until they reach the service stations. In particular, we propose to reduce the problem of coordinating the recharging of mobile sensors to the problem of finding energy-aware routes in a logical Compass Directed Unit Graph (CDG), defined below, built on top of the original topology. The proposed graph incorporates ideas from forward progress routing techniques [40–44] and the directionality of compass routing [45] in an energy-aware unit sub-graph.

Definition 6.1. Given the set of sensors S in $\mathbb{R}^2$, its Unit Disk Graph (or Unit Graph) is the graph $G = (S, E)$ where $\forall s_i, s_j \in S$, $(s_i, s_j) \in E$ if and only if $d(s_i, s_j) \leq R$, where d denotes the Euclidean distance and R is the transmission range.

Definition 6.2. Let $G = (S, E)$ be the unit disk graph of S, and let F be a recharge facility in $\mathbb{R}^2$. The Compass Directed unit Graph (CDG) of S with respect to F is the directed graph $G' = (V', E')$, where $V' = S \cup F$ and $\forall\ v_i, v_j \in V'$, $\overrightarrow{(v_i, v_j)} \in E'$ if and only if the following conditions are satisfied:

(1) Unit graph criterion: $d(v_i, v_j) \leq R$.
(2) Proximity criterion: $d(v_j, F) < d(v_i, F)$ and $d(v_i, v_j) < d(v_i, F)$
(3) Directionality criterion: $\exists v_{jp}$ such that $v_j \vec{v}_{jp} \cdot v_i \vec{F} = 0$ and $d(v_i, v_{jp}) + d(v_{jp}, F) = d(v_i, F)$.

Routing algorithms use the hop count as the metric to measure effectiveness. In our case, the hop count would be equivalent to the number of swapping operations between sensors in our CDG. Our solution to the FFP can be divided into two main stages: 1) the construction of the CDG and 2) the incremental swapping approach (i.e., migration) towards the rendezvous location.

6.4.2 *Creating the CDG*

An example of the proposed CDG for three sensors A,B,C and a facility F is shown in Figure 6.3. In this first stage, it is assumed that all sensors have the required levels of energy to construct the CDG. The process is rather simple and can be summarized by the following actions:

(1) Sensors position themselves at some initial fixed location that depends on the task application.
(2) Sensor A sends a NEIGHBOR_REQUEST broadcast message inviting other sensors to participate.

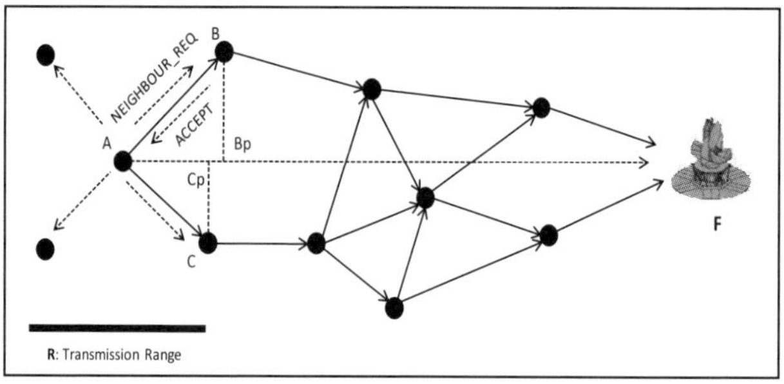

Fig. 6.3 Compass Directed Graph.

(3) Upon receiving a NEIGHBOR_REQUEST message from sensor A, immediate neighbors verify the neighboring criteria according to the following rules:
a) Proximity: $d(A, F) > d(B, F)$ and $d(A, B) < d(A, F)$.
b) Directionality: For example, B and C are neighbors of A if the corresponding projections B_p and C_p on line $\overline{AF}$ intersect the line segment $\overline{AF}$.

(4) If both conditions a) and b) are met, then sensors B and C send a NEIGHBOR_ACCEPT message. Otherwise they send a NEIGHBOR_DENY message.

In order to save energy, sensor A will then try to deviate as little as possible from the direction of the recharge station F. That is, sensor A will try to minimize the angle $\angle BAB_p$. Therefore, all the sensors that satisfy the conditions a) and b) are ranked according to the following function: $f(s_i, s_j) = \left\{ d(s_i, s_j) + \frac{d(s_j, s_{jp})}{d(s_i, s_j)} \right\}$ where s_i, s_j are the neighboring sensors, d is the Euclidean distance, F is the recharge station and s_{jp} is the projection of s_j on the line segment $\overline{s_i F}$.

At the end of this phase, each sensor will have two routing tables: one containing its children (i.e., sensors from which NEIGHBOR_ACCEPT messages were received) with their corresponding rank and a second table containing its parents (i.e., sensors to which NEIGHBOR_ACCEPT messages were sent). The routing tables are just partial maps of the network indicating the position of the children and parents.

6.4.3 *Migration strategy*

The second stage starts when sensors change their state from BATTERY_OK to BATTERY_LOW as a result of their battery levels falling below the first threshold. Once a sensor enters this state, it will try to get closer to the facility by making a series of one-hop swaps with its graph neighbors.

The swapping operation is initiated with a sensor sending a SWAP_REQUEST message to its lowest ranked neighbor. Neighbors could be ranked based on their distance (closest to farthest) and their direction relative to the target station. Another option of ranking includes the energy levels of neighbors as a metric as well as the energy levels of 2-hop neighbors (i.e., children of my children). If the current energy level of the child sensor is larger than the parent sensor, the sensor replies with a SWAP_ACCEPT message and travels to the position of the parent sensor. If its energy level is lower, it replies with a SWAP_DENY message. Once a requesting sensor has initiated the swapping process, it will not entertain any SWAP_REQUEST messages until the swapping operation is completed. The swapping operation is considered atomic and, once completed, both sensors will send a SWAP_COMPLETE message that will be used by current and new neighbors/parents to update their routing tables.

The final step of this phase takes place when battery levels fall enough to trigger a change to the BATTERY_CRITICAL state. In this state, the sensors behave exactly as in the passive approach and their interaction with the service station is defined by the pattern discussed earlier. A BATTERY_CRITICAL sensor sends a RECHARGE_REQUEST message to the recharge station and waits until an available socket is assigned. Similar to the passive approach, there are two cases to consider: 1) The recharge station is within the sensor's transmission range and 2) The recharging station is outside the sensor's transmission range and lowest ranked neighbors will forward the request towards the station. If there is no routing mechanism in place, the sensor can initiate its journey (i.e., panic situation) until the station is within range.

In an ideal system, all sensors will reach the BATTERY_CRITICAL state when they are exactly at one-hop distance from the recharge station. When the trip to the recharge station is made from a one-hop position in the graph (i.e., there are no graph neighbors), we call this "*one-hop run*" or "*optimal run*". Contrarily, if the trip is made from any other location, it is called a "*panic run*". We will come back to visit this issue when we discuss the experimental analysis of the different strategies.

There are two important properties of the CDG (i.e., dynamic and self-correcting) that can be explained by the following scenarios. Both scenarios may cause situations where the information in the neighboring tables is obsolete.

- Scenario 1: Simultaneous swapping. As part of the swapping process, the participating sensors exchange their neighboring information, that is, their corresponding children and parent tables. However, since multiple swapping operations may occur at the same time, when a sensor finally arrives at the position occupied by its swapping partner, the information in its neighboring tables may be out-of-date.
- Scenario 2: Sensor recharging. While this process takes place, other sensors may be swapping positions. Once the recharging process is finished, the sensor returns to its last known position. However, the structure of the network around it has changed. This situation is even more evident when trips to the facility are made from distances of more than one hop as a result of "panic runs".

The solution to these problems is to define the neighboring information as position-based tables, where the important factor is the relative position of the neighbors and not their corresponding IDs. The information of the actual sensors occupying the positions is dynamic. In other words, a sensor in a given position (x, y) knows that at any given point in time it has n children at positions $(x_1, y_1)...(x_n, y_n)$ and p parents at positions $(x'_1, y'_1)...(x'_p, y'_p)$. This information is static with respect to (x, y) and will not be modified. However, the identity of the sensors occupying the positions is dynamic and will get updated every time a swapping operation occurs. The mechanism to detect changes in the routing tables is triggered by sending a SWAP_COMPLETE message. When two neighboring sensors successfully complete a swapping operation, they will announce their new positions by sending SWAP_COMPLETE messages. Sensors within the transmission range that listen to this message will verify whether any of the positions involved in the exchange belong to their routing tables and update the appropriate entry with the ID of the new occupant of that position (i.e., self-correcting property).

On the other hand, a sensor returning from the service station (e.g., scenario 2) needs to re-discover the new occupants of its routing tables. This process is initiated by a SENSOR_RECHARGED message sent by the newly recharged sensor as soon it reaches its last occupied position on the network. Potential children and parents, upon receiving this message, will reply with CHILD_UPDATE and PARENT_UPDATE messages accordingly. This process is also used for parents to update their information about the energy levels of this newly recharged sensor.

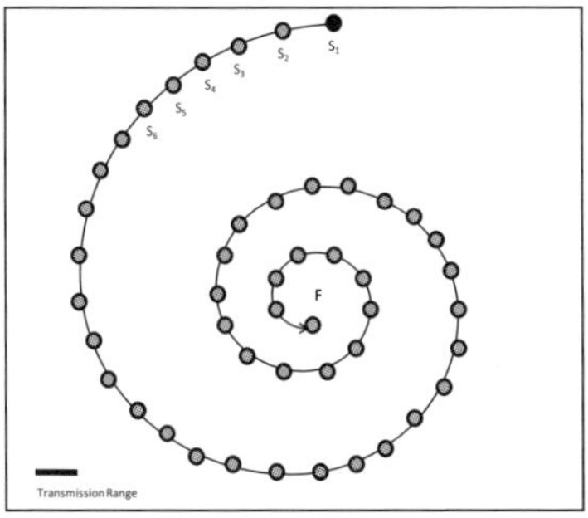

Fig. 6.4 Proactive strategy for Sensor s_1 in a spiral deployment.

These two important properties, along with a neighboring criteria that incorporates ideas from forward progress and compass routing [40, 41, 45] in an energy-aware unit graph, ensure the following lemma:

Lemma 6.1. *The swapping-based proactive solution to the FFP guarantees that all sensors reach the rendezvous location within a finite number of swapping operations.*

Proof. Let $G' = (V', E')$ be the CDG of S with respect to the recharge facility F. to prove the lemma, it suffices to show that every path $P =< s_{i,1}, ..., s_{i,k}, F >$ in G' from $s_{i,1}$ to the recharge station F does not contain any cycles; that is all $s \in P$ are distinct.

By contradiction, let $s_{i,j} = s_{i,r}, j < r$; this means that $s_{i,(j+1)}$ is a child in G' of sensor $s_{i,r}$, which means that $d(s_{i,(j+1)}, F) < d(s_{i,r}, F)$. This contradicts the proximity criterion (triangular inequality). Hence, the Lemma holds. □

6.4.4 *Extreme cases*

So far, the proactive strategy seems not only possible but intuitively more efficient than a passive approach. However, for some specific deployments, the proactive solution may not report any improvements over the passive approach. The deployment shown in Figure 6.4 shows the trajectory followed by a sensor s_1 during its migration towards the facility F.

If a proactive strategy is selected for this particular deployment, the sensor will start a gradual approach towards the facility. The sensors' limited transmission range implies that only one neighbor will be discovered during the CDG creation. That is, $\forall$ sensor $s_i \in \{s_1, s_2, ...\}$, s_{i+1} is graph neighbor (i.e., child node) of s_i (i.e., $d(s_{i+1}, F) < d(s_i, F)$) and s_{i-1} its corresponding parent. The locally-based swapping selection criteria will force sensor s_1 to exchange positions with its only available graph neighbor s_2. Consequently, s_1 will take the longest possible path to rendezvous with the facility.

In this particular example, the proactive strategy incurs an excessive and unnecessary waste of energy by a continuous sensor swapping. The sensor will eventually reach a BATTERY_CRITICAL state and will default to a passive behavior. However, this could have been avoided by taking a passive approach and waiting in its original position until the BATTERY_CRITICAL state is reached. In this particular deployment following a passive approach would have maximized sensing time by avoiding temporary coverage holes due to unnecessary swapping operations.

6.5 Improving the Proactive Strategy

The number of graph neighbors has a direct impact on the performance of the proactive strategy. Having more immediate graph neighbors implies more options when exploring a greedy migration towards the recharge station but it also means more interactions, notifications, etc., as more sensors will be affected by SWAP_COMPLETE and SENSOR_RECHARGE messages. Therefore, it may be beneficial to be more selective when choosing the graph neighbors and perhaps having fewer but better selected neighbors. The problem is to determine the right number of sensors within range that should be selected as graph neighbors. Here, we have a clear trade-off between flexibility when choosing a migration path and the required maintenance overhead.

So far, low battery sensors choose their swapping partners based on the energy levels of their 1-hop graph neighbors. Enhancing sensor knowledge by adding information about the energy levels of the 2-hop graph neighbors may impact the path selection process and facilitate the migration through higher energy areas of the network. However, is more knowledge better to achieve energy equilibrium? or is this added knowledge too costly?

6.5.1 *Exploring different topologies*

In this section we evaluate the same swapping-based migration strategy presented in Section 6.4.3 on a different underlying topology. The requirements for the new topology remain the same, and they are: 1) It should be built using local information only. 2) It should be flexible enough to operate in an asynchronous environment. 3) It should be dynamic, self-correcting. 4) Mobility strategies based on this topology should be loop-free.

For the new topologies to consider, the sensors will select their graph neighbors based on the concept of Gabriel neighbors and Relative neighbors. Two points A and B are said to be Gabriel neighbors if their diametric circle does not contain any other points. A graph where all pairs of Gabriel neighbors are connected with an edge is called the Gabriel graph. In our case, two sensors s_1 and s_2 with coordinates (x_1, y_1) and (x_2, y_2) are Gabriel neighbors if the circle with center $(\frac{x_1+y_1}{2}, \frac{y_1+y_2}{2})$ and radius $\frac{d(s_1,s_2)}{2}$ does not contain any other sensor. A particular case of a Gabriel Graph is the Relative Neighbor Graph where sensors s_1 and s_2 are relative neighbors if there are no other sensors in the *Lune* between sensors s_1 and s_2. That is, if $\forall\ S$, $S \neq s_1$ and $S \neq s_2$, $d(s_1, s_2) < max\,\{d(s_1, S), d(s_2, S)\}$ where d denotes the Euclidean distance between two sensors [46, 47].

In this new scenario, where low energy sensors will select their Gabriel or Relative neighbors as the potential swapping partners, the migration strategy towards the recharge station will be based on finding energy efficient routes on a Compass Directed Gabriel Graph (CDGG) or a Compass Directed Relative Neighbor Graph (CDRNG).

Definition 6.3. Let $G' = (S \cup F, E')$ be the compass directed unit graph of S with respect to recharge facility F. The *Compass Directed Gabriel Graph* (CDGG) of S with respect to F is the subgraph $\hat{G} = (S \cup F, \hat{E})$ of G' where, $\forall (s_i, s_j) \in E'$, $\overrightarrow{(s_i, s_j)} \in \hat{E}$ if and only if $\nexists s_k \in S$ such that $d(s_k, \frac{s_i+s_j}{2}) < d(s_i, \frac{s_i+s_j}{2})$.

6.5.2 *Creating the CDGG and CDRNG*

Figure 6.5 shows an example of the proposed CDGG for three sensors A,B,C and a facility F. In the first stage of the algorithm, it is assumed that all sensors have the required levels of energy to construct the CDGG. The process is similar to the creation of the CDG presented in Section 6.4.2. However, to guarantee that only the Gabriel neighbors are selected as graph neighbors, the sensor should implement the following actions:

(1) Upon reception of a NEIGHBOR_ACCEPT message from a potential Gabriel neighbor S', the receiving sensor S verifies if there is already

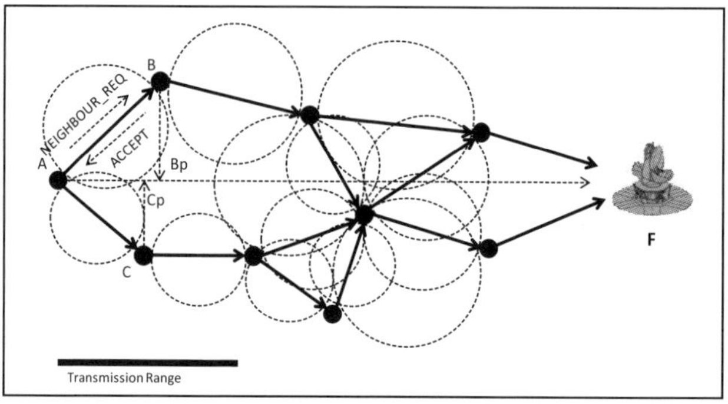

Fig. 6.5 Compass Directed Gabriel Graph.

a graph neighbor in the disk with center $(\frac{S_x+S'_x}{2}, \frac{S_y+S'_y}{2})$ and radius $\frac{d(S,S')}{2}$. If such a neighbor exists, then sensor S sends a NEIGHBOR_DENY message to S'.

(2) If no existing graph neighbor is found in the previous step, this means that sensor S' is in fact a Gabriel neighbor. However, some of the existing graph neighbors could be affected by this newly accepted sensor and they are no longer Gabriel neighbors. If the newly accepted sensor S' falls in the diametric disk between sensor S and one of the existing graph neighbors S_i, the neighbor in question should be excluded by sending it a NEIGHBOR_DENY message.

6.5.3 *Increasing sensor knowledge*

Another possible enhancement to improve the overall performance of the proactive strategy and help low energy sensors reach the recharge station faster is to add additional information about the energy levels of the 2-hop graph neighbors. Regardless of the topology chosen (i.e., CDG, CDGG, or CDRNG), having the 2-hop neighboring information combined with the 1-hop greedy strategy should lead to a more energy efficient path selection. To implement this new approach, a series of changes to the existing algorithms is necessary. For example, the neighboring information stored by each sensor s needs to change to include the tuple $(s_i, E_{S_i}, E_{S_{i_{2hop}}})$ where s_i is the *i-th* 1-hop neighbor of s. E_{S_i} represents the energy level and $E_{S_{i_{2hop}}}$ represents the average energy levels of the 1-hop graph neighbors of s_i.

The information about existing 1-hop graph neighbors will be appended to the NEIGHBOR_ACCEPT messages sent during the graph creation

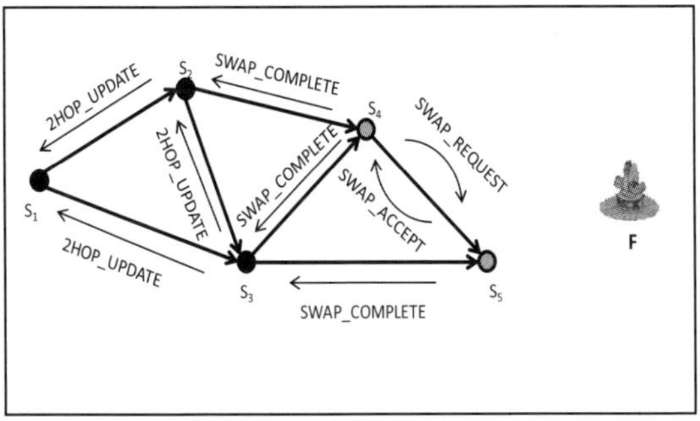

Fig. 6.6 Sensor swapping with 2-hop neighbors updates.

phase. When a sensor sends a NEIGHBOR_ACCEPT message to its parent, the message will now include the average energy level of its existing 1-hop neighbors. This new piece of information will have to be updated once the migration or swapping phase is initiated. Consequently, two swapping sensors will exchange this new piece of information as part of the swapping process. Furthermore, sensors reacting to a SWAP_COMPLETE message will generate a new message NEIGHBOR_2HOP_UPDATE to inform their parents about the changes of their 2-hop graph neighbors.

Let us examine the example shown in Figure 6.6 to illustrate the new interactions required during a swapping operation. In this example, sensors S_4 and S_5 have agreed to swap positions after the corresponding exchange of SWAP_REQUEST and SWAP_ACCEPT messages. Once the sensors arrive at the location occupied by their swapping partners, both sensors (i.e., S_4 and S_5) will send SWAP_COMPLETE messages to their parents S_2 and S_3. The SWAP_COMPLETE message received by sensor S_2 contains the tuple $(S_4, E_{S_4}, E_{S_{4_{2hop}}})$. After updating its neighboring information with the newly received information, S_2 computes the combined energy level of its 1-hop graph neighbors: $E_{S_{2_{2hop}}} = \frac{E_{S_3}+E_{S_4}}{2}$ and sends a new NEIGHBOR_2HOP_UPDATE $(S_2, E_{S_4}, E_{S_{2_{2hop}}})$ message to its parent S_1.

It is clear from the previous example that for each successful swapping operation there will be an overhead produced by the new NEIGHBOR_2HOP_UPDATE messages. The density of the graph, determined by the neighbor selection criteria and the sensor transmission ranges, will have a great impact on how many of these new notification messages are generated. The next section examines the impact of this added knowledge, its relationship with the underlying topology chosen, its potential benefits and possible drawbacks.

6.6 Experimental Results

Previous work on energy consumption of wireless sensor networks and protocols such as 802.11, have found that the energy required to initiate communication is not negligible. In particular, loss of energy due to retransmissions, collisions and acknowledgments is significant [48, 49]. Protocols that rely on periodic probe messages and acknowledgments are considered high cost. It is also noted in the literature that sensors' energy consumption in an idle state can be as large as the energy used when receiving data [49]. On the other hand, the energy used in transmitting data could be between $30 - 50\%$ more than the energy needed to receive a packet.

A common consideration for any solution involving mobile entities is how to accurately represent the energy spent when moving from one location to another. Locomotion cost depends on many factors such as the weight of the electronic components, irregularities in the terrain, obstacles, etc. For simplicity, in [26, 33], the weighted Euclidean distance between origin and destination is used as the cost of relocating a robot. In particular, in [33] is observed that the energy required to move their robotic sensors was 54 times the energy required to send a packet over the same distance and the energy spent in communications (i.e., send/receive) was 25% more than the battery drain in the idle state.

6.6.1 *Experimental environment and performance criteria*

The different scenarios are implemented in Omnet++ [50] along with the mobility framework extension [51]. For all experiments, the sensors and charging facilities were randomly placed in an area of $1000 \times 1000m^2$. The analysis of our simulated results centers on two important aspects of the solutions: 1) Whether or not a state of equilibrium is achieved and the number of sensor losses until such a condition is met; 2) Impact of several variables such as: underlying topology, transmission range, number of recharge sockets/ports and sensor knowledge.

In an ideal system, all sensors will reach the BATTERY_CRITICAL state when they are exactly at one-hop distance from the rendezvous location. When the trip to the recharge station is made from a one-hop position (i.e., there are no graph neighbors), it is called a "one-hop run" or "optimal run". Contrarily, if the final trip is made from any other location, it is called a "panic run" [32]. In all the simulated scenarios, the quality of the strategy is measured in terms of optimal runs vs. panic runs. Constant

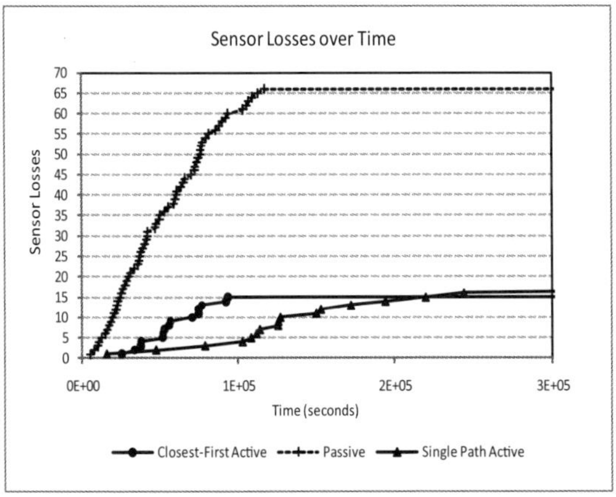

Fig. 6.7 Passive Strategy vs. Proactive Strategies.

cost values are assigned to each basic operation (i.e., send, receive, idle and move). Initial values for these operations are based on some of the observations found in [33, 48, 49].

6.6.2 *Passive vs. proactive*

The first set of experiments attempt to find out how the proactive strategies perform when compared to a passive approach. The goal is to measure the number of sensor losses due to battery depletion over time until the system reaches a state where no further sensor losses are reported (i.e., state of energy equilibrium). In this context, several proactive strategies are examined: 1) The closest-first strategy, where sensors attempt to make forward progress by swapping positions with the closest neighbor and 2) Single path strategy, where sensors select a single graph neighbor (e.g., first discovered).

The results of an experiment involving 100 sensors and one service facility are shown in Figure 6.7. The facility is equipped with two sockets, allowing two sensors to be recharged simultaneously. A series of 30 tests with different random deployments are run for 10^6 simulation seconds. The sensor transmission range is fixed at 100m and the energy ratio for sending/receiving a packet is set to a constant ($E : E/2$). Locomotion costs were based on the weighted Euclidean distance with a weight factor of $\frac{1}{5}E$ per meter traveled. The results show that the two variations of the proactive strategy reached the state of equilibrium. This means that all the

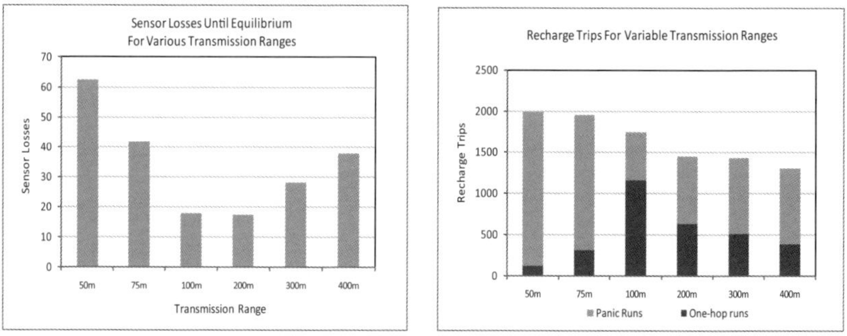

(a) (b)

Fig. 6.8 Experimental results: Variable Range.

energy spent during the graph creation, swapping and graph reconfiguration in a network with a 100:1 sensor-facility ratio with two sockets did not overwhelm the system to the point of preventing it from reaching equilibrium. Surprisingly, even the single path proactive strategy outperformed the passive approach by a significant margin.

In comparison, for a similar network size, the solutions presented in [11] and [19] required 2 and 3 stations, or actors, respectively, to maintain a "live" network (i.e., 50% or more sensors remain after equilibrium was reached). In our case, equilibrium was achieved with 1 facility with two docking ports for a similar network size and over 80% of network survivability.

Even though the passive strategy reached the state of equilibrium faster than the single-path proactive strategy, the cost in terms of sensor losses was very high. This result implies that if a passive approach is chosen for high sensor-facility ratio deployments, the number of recharge sockets in this experiment is too restrictive. This result is similar to the passive approach followed in [19] where it was observed a significant improvement by adding a second recharge station.

6.6.3 *Transmission range*

This experiment was designed to verify the impact of the sensor's transmission range on the overall performance. The characteristics of the network were the same as the test performed in Section 6.6.2 with the closest-first swapping strategy. The only difference is that the transmission range was varied from 50m, 75m, 100m, 200m, 300m to 400m.

Figure 6.8(a) shows the cumulative number of sensor losses until equilibrium for each range value. In a deployment of 1000x1000m^2 a transmission range of 50m was too restrictive, which means that most of the sensors were isolated and the number of immediate neighbors in the CDG was too small to guarantee a gradual approach towards the recharge location. Another interesting observation is that by increasing the transmission range, the number of losses decreased dramatically. However, for larger ranges (e.g., 300m and 400m), there was a decline on the overall performance since many neighbors are discovered, resulting in an added overhead to maintain more information per sensor as well as additional interactions due to update messages as a result of successful swapping and recharging operations.

Figure 6.8(b) shows the quality of the solution in terms of one-hop runs vs. panic runs. In an ideal system, our solution should reach the state of equilibrium using one-hop runs only. As expected, for a transmission range of 50m, most of the trips could be considered panic runs since there is almost no migration due to the lack of 1-hop neighbors. The best breakdown between one-hop and panic runs occurs with 100m range. However, there are more visits to the recharge location, when compared to the 200m, 300m and 400m cases. Although there is no clear explanation for this phenomenon, one can argue that there is a trade-off between the total number of recharge trips and the breakdown between one-hop vs. panic runs. In a panic run situation, a sensor travels from a more distant location and after having been recharged, it needs to return farther to its initial location. This situation creates a coverage hole that lasts longer than holes created by one-hop runs. However, more one-hop recharge trips also means more coverage holes but for shorter periods of time.

6.6.4 *Topology comparison*

This test was designed to determine whether our proactive solution to energy restoration reaches a state of equilibrium when the new proposed CDGG and CDRNG are used as the underlying topologies for the mobility strategies. The experiment measured the cumulative number of sensor losses until energy equilibrium is reached.

Figure 6.9(a) shows the result of simulations on the same network (see Section 6.6.2) involving 100 sensors with fixed transmission range of 100m and one service facility. For all the tests performed on the three different topologies, the mobility strategy selected was the greedy closest-first swapping where a low energy sensor chooses its closest graph neighbor as a swapping partner during its migration towards the recharge station.

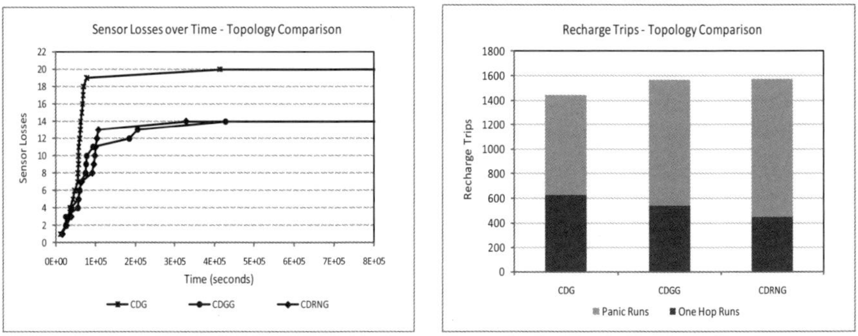

(a) (b)

Fig. 6.9 Experimental results: Topology Comparison.

As expected, the closest-first swapping strategy on the three topologies chosen (i.e., CDG, CDGG and CDRNG) reached the state of equilibrium. The CDGG and CDRNG are sub-graphs of the CDG and according to the experimental results presented in Section 6.6.2, even the single path (i.e., single neighbor) approach reached the state of equilibrium. However, the interesting finding is that although the three topologies reached the state of equilibrium at the same time approximately, the CDGG and CDRNG reported fewer sensor losses due to battery depletion. This is an important observation that implies that fewer but better selected graph neighbors will yield better results if the main goal is to minimize the number of permanent failures due to battery depletion.

Unfortunately, the CDGG and CDRNG did not report any improvements in terms of optimal trips to the recharge station. Figure 6.9(b) shows the number of recharge trips and breakdown between optimal and panic runs for the three topologies in question. For the CDGG and CDRNG there was a small increase in the number of recharge visits compared to the CDG and a small decrease in the number of optimal runs. This decrease is somehow expected since the number of neighbors for both topologies (i.e., CDGG and CDRNG) is more restrictive than the CDG. Once more, choosing different topologies for the migration strategy exposed a trade-off between permanent coverage holes due to battery depletion and more short-lived temporary holes due to more frequent visits to the facility.

The next part of this test was designed to measure the impact of the recharge sockets on the cumulative number of losses until equilibrium and verify whether the perfect equilibrium can be reached by increasing the

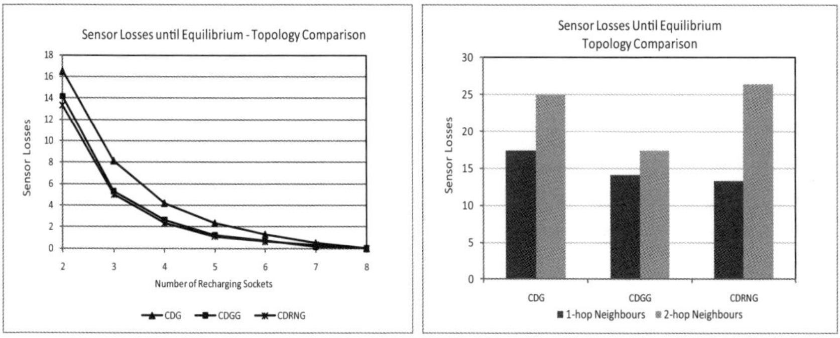

(a) (b)

Fig. 6.10 Experimental results: Topology Comparison II.

number of sockets or docking ports in the recharge station. The network setup remained the same and the closest-first greedy mobility strategy was tested on the three topologies (i.e., CDG, CDGG and CDRNG). Figure 6.10(a) shows the result for this test where the closest-first swapping strategy on the three topologies showed the same progression towards perfect equilibrium. The total number of recharge sockets needed for the perfect equilibrium is the same for the three topologies but the CDGG and CDRNG showed an improvement on the number of sensor losses over the CDG as the number of recharge sockets increased.

6.6.5 *Sensor knowledge*

The goal of this set of tests is to verify the impact of added sensor knowledge, as introduced in Section 6.5.3, and compare it with the 1-hop information greedy strategies on the three proposed topologies. The network parameters are the same as in the previous tests, with fixed transmission range at 100m. The closest-first swapping strategy is applied on the three topologies (i.e., CDG, CDGG and CDRNG) with information about the energy levels of 1-hop graph neighbors only and 2-hop graph neighbors respectively.

Figure 6.10(b) shows the number of sensor losses until equilibrium for the three topologies tested with 1-hop neighbor information vs. 2-hop neighbor information. In each case, there was an increase in the number of sensor losses when the migration strategy included the 1-hop neighbor information. When 2-hop information is used, the best performer was the

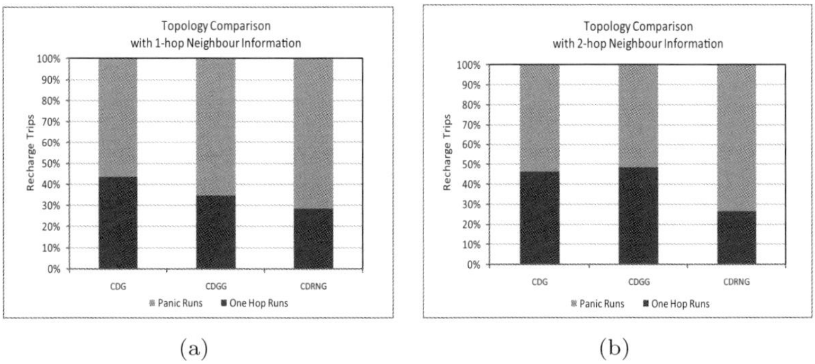

Fig. 6.11 Experimental results: Sensor Knowledge.

CDGG with losses similar to the 1-hop CDG. This is a rather surprising result, which seems to imply that "knowing more individually" about the network is less useful for the collective effort than "knowing less". Knowing more in this case has a direct impact on the number of control messages required to maintain the underlying topology in a consistent state. This phenomenon will be more evident as the graph degree increases. The graph maintenance overhead related to keeping 2-hop neighbor information proved to be crucial to the point that counteracts any possible improvement when compared to keeping 1-hop information only.

The idea of adding extra knowledge to the sensors aimed to improve the path selection strategy and increase the number of optimal runs or 1-hop trips to the recharge station. The simulation results shown in Figure 6.11(b) confirmed our expectations. Added knowledge had, in fact, a positive impact on the selection of the better energy-efficient migration strategy towards the recharge station. There was some marginal improvement on the number of optimal runs for the CDG and CDRNG with a real improvement for the CDGG. The CDGG proved again to be the best performing topology in terms of cumulative sensor losses until equilibrium and breakdown between panic and optimal runs when using 2-hop neighbor information.

6.7 Closing Remarks and Open Problems

Throughout this chapter we have focused on the problem of energy restoration in a mobile sensor network with static recharging facilities. The *perfect* energy restoration strategy should be able to guarantee a continuous operation of the network without any losses; however in reality some sensor losses will occur. A *successful* energy restoration strategy is one in which the losses are limited in scope and time; that is, the strategy allows the network to reach a state of equilibrium, where no further sensor losses will occur (thus guaranteeing a continuous operation of the network), with the minimum number of sensor losses.

We have examined the basic approaches, passive and proactive, to energy restoration strategies. To compare the quality of the approaches, passive vs proactive, and the difference between different proactive mobility strategies we focused our analysis on several key indicators, such as the number of sensor losses until equilibrium is reached, the distance traveled to reach the recharge station (i.e., optimal runs vs. panic runs), and the amount of resources needed to achieve a perfect equilibrium (i.e., without any loss ever). The analysis shows the definite advantages of taking a proactive approach to energy restoration.

In terms of proactive strategies, the problem of coordinating the recharging of mobile sensors has been reduced to the problem of finding optimal routes in a logical Compass Directed Graph (CDG) or Compass Directed Gabriel Graph (CDGG) built on top of the original deployment. All the proactive solutions analyzed here have three important properties [32, 33]: 1) The proposed graphs guarantee that sensors reach the recharge facilities in a finite number of swapping operations (the trajectory is loop-free). 2) All decisions made by the sensors regarding the next swapping operation are based on local knowledge (i.e., the algorithms are completely distributed and localized). 3) New sensors can be added or deleted at any time and new neighbors are re-discovered any time a successful swapping or recharge operation takes place, making the graphs dynamic and self-correcting.

Many important problems are open and need to be addressed. Some of them are listed in the following paragraphs.

The proactive strategies proposed in [32, 33] and examined in this Chapter assume that there are no *obstacles* between the sensors and the chosen recharge facility. However, more challenging environments may contain static obstacles that prevent the sensors from communicating with other sensors or traveling directly to the recharge station. The presence of obstacles in static sensor networks has been the focus of attention in several research papers. For example, in [52] a model for obstacles in static sensor networks is discussed, where obstacles are distinguished into physical and communication ones. A physical obstacle is a network area which prevents the deployment and movement of sensors in that area.

A communication obstacle, on the other hand, causes a disruption to the wireless communication: if the line of sight between two sensors crosses the obstacle, then there is no communication between those sensors. The problem of how to successfully route packets around obstacles has been examined in [53–55]. In mobile sensor networks, this problem has not yet been examined; clearly, the obstacle avoidance strategies have a higher degree of complexity since the algorithms have to guarantee not only communication but also movement around the obstacles.

There are several unexplored variants of the energy restoration problem. For example, let $P = \{p_1, ..., p_K\}$ a set of points of interests and f, a static recharge facility. In this variant, each sensor s must visit an assigned set of points $P_s \subset P$ and repeat this process continuously; in other words, each sensor has to visit a sub-set of points of interest continuously and also visit the facility periodically to recharge its battery. Within this variant, there a number of alternatives: 1) Ordered vs. non-ordered points of interests: the sensor has to visit the assigned set of points in a given order or in an arbitrary order (but always guaranteeing the all points are visited before starting the next round of visits). 2) Disjoint vs overlapping routes or itineraries: $P_{s_i} \cap P_{s_j} = \emptyset, \forall s_i, s_j \in S$ or $P_{s_i} \cap P_{s_j} \neq \emptyset$; in the latter case, two sensors must not visit the same point at the same time. 3) Fixed vs. exchangeable itineraries or point of interest: the assigned itineraries are fixed, or sensor can exchange their itineraries or "pick-up" other sensors' point of interest (e.g., if a sensor dies of energy starvation, another sensor can add the depleted sensor's itinerary to its own).

For all these variants, the main goal is the same as the original problem: to achieve a state of equilibrium where, without any further losses, the sensors fulfill their tasks but also cooperate to share a recharge station with limited resources (i.e., number of recharge sockets).

References

[1] G. Wang, G. Cao, T. L. Porta, and W. Zhang, Sensor relocation in mobile sensor networks, *In Proceedings of IEEE INFOCOM.* pp. 2302–2312 (2005).

[2] X. Li, N. Santoro, and I. Stojmenovic, Mesh-based sensor relocation for coverage maintenance in mobile sensor networks, *In Proceedings of the 4th international conference on Ubiquitous Intelligence and Computing.* pp. 696–708 (2007).

[3] J. Luo and J. P. Hubaux, Joint mobility and routing for lifetime elongation in wireless sensor networks, *In Proceedings of the 24th Annual Joint Conference of the IEEE Computer and Communications Societies (IEEE Infocom 2005).* pp. 1735–1746 (2005).

[4] W. Wang, V. Srinivasan, and K. Vikram, Extending the lifetime of wireless

sensor networks through mobile relays, *IEEE/ACM Transactions on Networking.* **16**, 1108–1120 (2008).

[5] M. Rahimi, H. Shah, G. Sukhatme, J. Heidemann, and D. Estrin, Studying the feasibility of energy harvesting in a mobile sensor network., *In Proceedings of the IEEE International Conference on Robotics and Automation.* pp. 19–24 (2003).

[6] S. Roundy, P. Otis, Y. Chee, J. Rabaey, and P. Wright, A 1.9ghz rf transmit beacon using environmentally scavenged energy., *Digest IEEE International Symposium on Low Power Electricity and Devices* (2003).

[7] X. Li, A. Nayak, and I. Stojmenovic, Exploiting actuator mobility for energy-efficient data collection in delay-tolerant wireless sensor networks., *In Proceedings of the Fifth International Conference on Networking and Services ICNS.* pp. 216–221 (2009).

[8] Y. Mei, C. Xian, S. Das, Y. Hu, and Y. Lu, Sensor replacement using mobile robots., *Computer Communications.* **30**, 2615–2626 (2007).

[9] D. Simplot-Ryl, I. Stojmenovic, and J. Wu, Energy efficient backbone construction, broadcasting, and area coverage in sensor networks., *Chapter 11. Handbook of Sensor Networks: Algorithms and Architectures.* pp. 343–379 (2005).

[10] T. Tirta, B. Lau, N. Malhotra, S. Bagchi, L. Z., and Y. Lu, Controlled mobility for efficient data gathering in sensor networks with passively mobile robots., *Sensor Network Operations by Wiley-IEEE Press.* (2005).

[11] M. Sharifi, S. Sedighian, and M. Kamali, Recharging sensor nodes using implicit actor coordination in wireless sensor actor networks, *In Wireless Sensor Network.* **2**, 123–128 (2010).

[12] W. Heinzelman, A. Chandrakasan, and H. Balakrishnan, An application-specific protocol architecture for wireless microsensor networks, *IEEE Transactions on Wireless Communications.* **1**(4), 660–669 (2002).

[13] C.-H. Lung, C. Zhou, and Y. Yang, Applying hierarchical agglomerative clustering to wireless sensor networks, *In Proceedings of the International Workshop on Theoretical and Algorithmic Aspects of Sensor and Ad-hoc Networks (WTASA).* pp. 97–105 (2007).

[14] M. Perillo, C. Zhao, and W. Heinzelman, On the problem of unbalanced load distribution in wireless sensor networks, *In Proccedings of the Global Telecommunications Conference Workshops (GlobeCom 2004).* pp. 74–79 (2004).

[15] O. Younis and S. Fahmy, Heed: A hybrid, energy-efficient, distributed clustering approach for ad hoc sensor networks, *IEEE Transactions on Mobile Computing.* **3**, 366–379 (2004).

[16] F. Michaud and E. Robichaud, Sharing charging stations for long-term activity of autonomous robots, *In Proceedings of IEEE/RSJ International Conference on Intelligent Robots and Systems.* **3**, 2746–2751 (2002).

[17] J. Warwerla and R. Vaughan, Near-optimal mobile robot recharging with the rate-maximizing forager, *In Proceedings of the 9th European Conference on Artificial Life.* pp. 776–785 (2007).

[18] F. Arwin, K. Samsudin, and A. R. Ramli, Swarm robots long term autonomy

using moveable charger, *In Proceedings of the 2009 International Conference on Future Computer and Communication.* pp. 127–130 (2009).

[19] A. Drenner and N. Papanikolopoulos, Docking station relocation for maximizing longevity of distributed robotic teams, *In Proceedings of the 2006 IEEE International Conference on Robotics and Automation.* pp. 2436–2441 (2006).

[20] A. Howard, M. J. Mataric, and G. S. Sukhatme, Mobile sensor network deployment using potential fields: A distributed, scalable solution to the area coverage problem, *In Proceedings of DARS.* pp. 299–308 (2002).

[21] G. Wang, G. Cao, and T. Porta, Movement-assisted sensor deployment, *IEEE Trans. Mobile Computing.* **5**, 640–652 (2006).

[22] X. Li, H. Frey, N. Santoro, and I. Stojmenovic, Strictly localized sensor self-deployment for optimal focused coverage, *IEEE Transactions on Mobile Computing.* **10**, 1520 – 1533 (2011).

[23] Y. Toksoz, J. Redding, M. Michini, B. Michini, J. P. How, M. Vavrina, and J. Vian, Automated battery swap and recharge to enable persistent uav missions, *AIAA Infotech, Aerospace Conference* (2011).

[24] L.P.Koh and S. A. Wich, Dawn of drone ecology: low-cost autonomous aerial vehicles for conservation, *Tropical Conservation Science.* **5**, 121–132 (2012).

[25] P. McGillivary, K. Rajan, J. de Sousa, F. Leroy, and R. Martins, Integrating autonomous underwater vessels, surface vessels and aircraft as persistent surveillance components of ocean observing studies, *In Proceedings of the 2012 IEE/OES Autonomous Underwater Vehicles (AUV) Conference.* pp. 1–5 (2012).

[26] Y. Litus, R. Vaughan, and P. Zebrowski, The frugal feeding problem: energy-efficient, multi-robot, multi-place rendezvous, *In Proceedings of the 2007 IEEE International Conference on Robotics and Automation.* pp. 27–32 (2007).

[27] Y. Litus, P. Zebrowski, and R. T. Vaughan, A distributed heuristic for energy-efficient multirobot multiplace rendezvous, *IEEE Transactions on Robotics.* **25**, 130–135 (2009).

[28] A. Couture-Beil and R. Vaughan, Adaptive mobile charging stations for multi-robot systems, *In Proceedings of the 2009 IEEE/RSJ International Conference on Intelligent Robots and Systems.* pp. 1363–1368 (2009).

[29] M. I. Afzar, W. Mahmood, and A. H. Akbar, A battery recharge model for wsns using free-space optics (fso), *In Proceedings of the 12th IEEE International Multitopic Conference.* pp. 272–277 (2008).

[30] M. I. Afzar, W. Mahmood, S. M. Sajid, and S. Seoyong, Optical wireless communication and recharging mechanism of wireless sensor network by using ccrs, *International Journal of Advance Science and Technology.* **13**, 59–68 (2009).

[31] Y. Peng, Z. Li, W. Zhang, and D. Qiao, Prolonging sensor network lifetime through wireless charging, *In Proceedings of the 31st IEEE Symposium on Real-Time Systems (RTSS).* pp. 129–139 (2010).

[32] E. Velazquez and N. Santoro, Mobility-based strategies for energy restoration in wireless sensor networks, *In Proceedings of the 6th International Conference on Mobile Ah-hoc and Sensor Networks (MSN 2010).* pp. 161–168 (2010).

[33] E. Velazquez, N. Santoro, and M. Lanthier, Pro-active strategies for the frugal feeding problem in wireless sensor networks, *In Proceedings of the 2nd International ICST Conference on Sensor Systems and Software (S-CUBE 2010)*. pp. 189–204 (2010).

[34] P. Hebden and A. Pearce, Distributed asynchronous clustering for self-organisation of wireless sensor networks, *In Proceesings of the 4th International Conference on Intelligent Sensing and Information Processing (ICISIP 2006)*. pp. 37–42 (2006).

[35] L. Tan, Y. Gong, and G. Chen, A balanced parallel clustering protocol for wireless sensor networks using k-means techniques, *In Proceedings of the 2nd International Conference on Sensor Technologies and Applications (SENSORCOMM 2008)*. pp. 300–305 (2008).

[36] K. Xu, Y. Jia, and Y. Liu, A novel hierarchical clustering routing algorithm for wireless sensor networks, *In Proceedings of the 2008 International Conference on Internet Computing in Science and Engineering*. pp. 282–285 (2008).

[37] L. X. Zhang, An efficient energy adaptive clustering leach in wireless sensor network, *Key Engineering Materials*. **439**, 510–515 (2010).

[38] C.-H. Lung and C. Zhou, Using hierarchical agglomerative clustering in wireless sensor networks: An energy-efficient and flexible approach, *Ad Hoc Networks*. **8**(3), 328–344 (2010).

[39] G. Wang, G. Cao, and T. L. Porta, Movement-assisted sensor deployment, *IEEE Transactions on Mobile Computing*. **5**(6), 640–562 (2006).

[40] H. Frey, S. Ruhrup, and I. Stojmenovic, Routing in wireless sensor networks, *Chapter4. Guide to Wireless Sensor Networks*. pp. 81–111 (2009).

[41] I. Stojmenovic and X. Lin, Power-aware localized routing in wireless networks, *IEEE Transactions on Parallel and Distributed Systems*. **12**, 1122–1133 (2001).

[42] T. Z.-J. ation, W. Yi, and G. Zheng-Hu, Eegfgr: An energy-efficient greedy-face geographic routing for wireless sensor, *In Proceedings of the 2007 IFIP international conference on Network and parallel computing*. **4672**, 171–182 (2007).

[43] S. Ruhrup, H. Kalosha, A. Nayak, and I. Stojmenovic, Message-efficient beaconless georouting with guaranteed delivery in wireless sensor, ad hoc, and actuator networks, *IEEE/ACM Transactions on Networking*. **18**, 95–108 (2010).

[44] M. Z. Zamalloa, K. Seada, B. Krishnamachari, and A. Helmi, Efficient geographic routing over lossy links in wireless sensor networks, *ACM Transactions on Sensor Networks*. **4**, 1–33 (2008).

[45] E. Kranakis, H. Singh, and J. Urrutia, Compass routing on geometric networks, *In Proceedings of the 11th Canadian Conference on Computational Geometry*. pp. 51–54 (1999).

[46] K. Supowit, The relative neighborhood graph, with an application to minimum spanning trees, *Journal of the ACM*. **30**, 428–448 (1983).

[47] J. W. Jaromczyk and G. T. Toussaint, Relative neighborhood graphs and their relatives, *In Proceedings of the IEEE*. **80**, 1502–1517 (1992).

[48] L. Feeney, An energy consumption model for performance analysis of routing protocols for mobile ad hoc networks, *Mobile Network Applications.* **6**, 239–249 (2001).

[49] L. Feeney and M. Nilsson, Investigating the energy consumption of a wireless network interface in an ad hoc networking environment, *In Proceedings of the 20th Annual Joint Conference of the IEEE Computer and Communications Societies (IEEE Infocom 2001).* **3**, 1548–1557 (2001).

[50] A. Vargas, The omnet++ discrete event simulation system., *In Proceedings of the European Simulation Multi-conference (ESM'2001).* pp. 319–324 (2001).

[51] W. Drytkiewicz, S. Sroka, and V. Handziski, A mobility framework for omnet++., *In Proceesings of the 3rd International OMNeT++ Workshop* (2003).

[52] I. Chatzigiannakis, G. Mylonas, and S. Nikoletseas, Modeling and evaluation of the effect of obstacles on the performance of wireless sensor networks, *In Proceedings of the 37th Annual Symposium on Simulation (ANSS'06).* pp. 50–60 (2006).

[53] P. Bose, P. Morin, I. Stojmenovic, and J. Urrutia, Routing with guaranteed delivery in ad hoc wireless networks, *Wireless Networks.* **7**, 609–616 (2001).

[54] C. Y. Chang, C. T. Chang, Y. C. Chen, and S. C. Lee, Active route-guiding protocols for resisting obstacles in wireless sensor networks, *IEEE Transactions on Vehicular Technology.* **59**, 4425–4442 (2010).

[55] E. Hamouda, N. Mitton, B. Pavkovic, and D. Simplot-Ryl, Energy-aware georouting with guaranteed delivery in wireless sensor networks with obstacles, *Journal of Wireless Information Networks.* **16**, 142–153 (2009).

Chapter 7

Deployment and Redeployment of Wireless Sensor Networks: a Swarm Robotics Perspective

Andreagiovanni Reina[1], Vito Trianni[2]

[1] *IRIDIA, Université Libre de Bruxelles, Brussels, Belgium,*
[2]*ISTC, Consiglio Nazionale delle Ricerche, Rome, Italy*

Abstract. *Deployment of WSNs is an important issue that requires careful consideration, as it can make the difference between an efficient and an unproductive system. The introduction of node mobility provides a wealth of potential solutions to the deployment problem, which can lead to higher robustness, flexibility and adaptivity. When provided with mobility features, network nodes are analogous to autonomous robots with local sensing and communication abilities. Therefore, behavioral strategies developed for collections of autonomous robots may be exploited in the mobile WSN domain. This is particularly true for swarm robotics studies, which emphasize self-organizing behaviors that deal with limited individual abilities, local sensing and local communication. In this chapter, we discuss the challenges and opportunities offered by swarm robotics with respect to the deployment of mobile WSNs. We review the state of the art in swarm robotics for coverage, exploration and navigation tasks, which are directly linked to the deployment problem, and we identify relevant directions for an hybridization of WSN and swarm robotics research.*

7.1 Introduction

The deployment of a wireless sensor network is a relevant issue that can have a strong impact on the system efficiency and on the quality of service. Speaking in general terms, any given application may have specific deployment requirements that have to be met to optimize the network performance, preventing the usage of a priori generic strategies for deployment, and requiring flexible and adaptable methods [44].

Fixed deployment strategies may be unpractical for the application domain or too costly, for instance in case of harsh environmental conditions that do not allow precise positioning of sensors. The requirements for deployment may also vary in time, either due to application-specific demands or to variability in the network operating conditions (e.g., failure of some nodes). As a consequence, research in WSN deployment has often postulated the need of mobile sensors capable of positioning themselves starting from a deployment site, or capable of repositioning after an initial coarse deployment (e.g., launch from a plane). In this context, a number of different studies can be found, in which distributed deployment algorithms are proposed to obtain a desired distribution of sensors in the environment [1, 2, 13, 20, 27, 41, 45].

As soon as mobility comes into the game, sensor nodes become more and more similar to autonomous agents that decide their motion according to the environmental contingencies they experience, both related to the mission (i.e., the phenomenon to be monitored) and the network itself (i.e., the neighboring nodes and their operating conditions). If we exclude minimalistic mobility features, such as spring-propelled nodes capable of a single flip [7], mobile nodes can be surely considered autonomous robots with their sensing and motion capabilities. Alternatively, autonomous robots can be exploited to move sensor nodes to a desired location [6, 26, 37]. In both cases, the particular context of mobile WSNs naturally lends itself to a parallel with swarm robotics systems. Indeed, mobile WSNs and swarm robotics systems share the same foundational characteristics: large number of nodes/agents, local sensing and communication, limited individual sensory-motor and processing capabilities. It is therefore interesting to look at the swarm robotics literature to draw a parallel between the approaches exploited in swarm robotics related to coverage, exploration and navigation, as they can give novel solutions for the WSN domain, either inspiring novel algorithms for mobile nodes, or by providing robotic solutions for the automatic deployment, replacement and redeployment of (immobile) sensor nodes. In this chapter, we will examine the swarm robotics literature and will draw a parallel among swarm robotics and WSN algorithms. In Section 7.2, we discuss the specific challenges and opportunities offered by swarm robotics, with particular reference to the problems faced in the context of WSNs. In Section 7.3, we review the state of the art in swarm robotics discussing the most advanced solutions for problems of coverage, exploration and navigation. In Section 7.4, we discuss possible applications of swarm robotics approaches to the WSN domain, as well as cross-fertilization between the two. Section 7.5 concludes the chapter.

7.2 Challenges and Opportunities Offered by Swarm Robotics

Swarm robotics is a specific approach in the design and control of distributed multi-robot systems, which is characterized by a strong emphasis on self-organization as the main way to obtain desired system properties, such as scalability, flexibility and robustness. As mentioned above, swarm robotics systems are relatively large in the number of robots, which goes from few tens to many hundreds. Each robot is completely autonomous in its control, and behaves according to simple rules of thumb based on local sensing and local communication with neighboring agents. Decentralization of control and locality of available information make the design of swarm robotic systems particularly challenging, because it is necessary to identify which can be the locally-executed control rules that will lead to a desired global behavior. More specifically, the design problem is created by the indirect, often non-linear relationship between the individual behavior and the swarm organization. This makes it difficult to predict what is the macroscopic effect of microscopic control rules usually performed only with partial and noisy information.

An important assumption that characterizes swarm robotics systems is the lack of absolute positioning information. Normally, robots do not mount GPS receivers to provide a location estimate. Neither other absolute localization and tracking techniques are envisaged for robot control (e.g., the Vicon system[1]). Indeed, swarm robotics systems are intended for unstructured environments where the absence of any infrastructure is compensated by the cooperative action of a large number of robots. In this perspective, potential applications are search and rescue in disaster areas, space exploration or underseas monitoring. The lack of absolute positioning represents the bigger difference between swarm robotics and the mobile WSN domain. In mobile WSNs, motion of the nodes is (locally) planned on the basis of an absolute reference frame and the absolute position of neighboring nodes, and distributed algorithms are deployed with predictable properties that are often based on the availability of positioning

[1] http://www.vicon.com

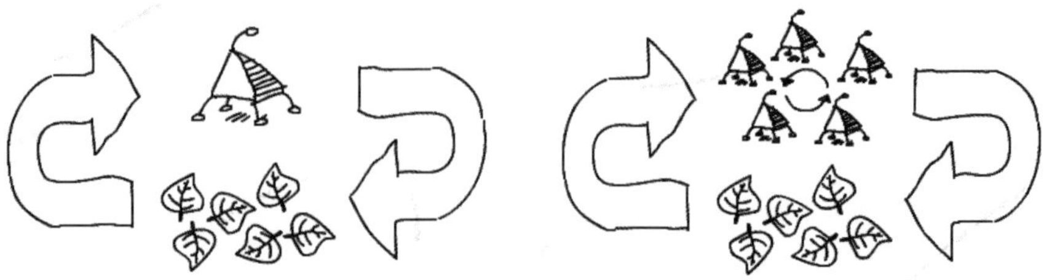

Fig. 7.1 A pictorial description of the sensory-motor loop. Left: a robot in interaction with the environment. In the single robot case, the action of the robot is executed in the environment, and the new environmental situation is perceived through the robot sensors. Right: in the swarm robotics case, multiple robots act at the same time and determine the state of the environment as it will be perceived by their sensors. Additionally, robots communicate and interact among themselves.

information. The challenge that is given by the relaxation of the localization assumption requires that the motion planning be performed in a local reference frame, on the basis of the communication with neighboring nodes. In these conditions, swarm robotics offers several algorithms that could be directly translated to the WSN domain, as we will see in Section 7.3.

Another assumption that is often made in WSNs and that is instead relaxed in swarm robotics systems concerns the maintenance of a global connectivity among the nodes. Indeed, disconnected nodes in a WSN are of no use for the application in place. This is why the maintenance of connectivity is one of the first properties to check in a distributed algorithm for mobile sensors. In swarm robotics, instead, the autonomy of the individual robots in executing their behavior allows them to continue working even when connection with other robots is (temporarily) lost. The higher mobility of robots with respect to mobile sensors and the possibility of dead-reckoning thanks to proprioceptive sensors (i.e., wheels encoders) make robotic systems more flexible and versatile. Still, when deployed in open environments and harsh environmental conditions (e.g., underseas), the maintenance of connectivity is an important challenge to be faced.

The ability to recover connectivity when lost is also supported by fast sensory-motor loops which normally characterize the autonomous robots behavior. In this respect, robots are different from mobile sensors in the fact that the behavior is the result of a fine-grained succession of actions and sensations. This is particularly true for reactive behaviors, such as those usually implemented in swarm robotics systems. In this case, the behavior performed by an individual robot often follows simple rules, that is, an action is chosen principally on the basis of the current sensory perception (plus some internal state information), and the executed action partially determines the following perceptual state in which the robot will find itself, therefore starting a novel sensory-motor loop (see Figure 7.1 left). Controlling a robot at such a fine-grained level is a challenging task, but this also offers the opportunity to exploit fast control loops to recover from errors and adapt to the environmental contingencies at hand. In the swarm robotics context, it is the whole swarm that rapidly changes state, and this has a bearing on the flow of perceptions that each individual robot experiences. Additionally, direct communication between robots must be taken into account to determine the action to be performed (see Figure 7.1 right). Therefore, controlling a robot within a swarm is challenging because the sensory state of a robot is not only determined by the action just performed, but it depends on the actions and state of the (possibly many) neighboring robots. An important challenge is however accompanied also by great opportunities, as the complexity of the behavior exhibited by a swarm goes far beyond the individual capabilities. A swarm can display quick information spreading and collective responses that are faster than the individual reaction times. It can take decisions without any single individual testing more than a single alternative. More generally, a swarm robotic system can behave optimally despite the inability of individual robots to acquire a global picture of the problem.

The above considerations summarize some of the most relevant differences between the domains of swarm robotics and of WSNs. Nevertheless, as mentioned in the introduction, it is worth making a parallel between these two domains, as cross-fertilization may provide a leap forward for both. In the following section, we therefore review the swarm robotics literature and discuss the relevant work that resonate with the problems faced in deployment and redeployment of mobile WSNs.

7.3 Current Approaches in Swarm Robotics

There is a large number of studies that can be considered in the attempt to look at the problems faced in WSNs from a swarm robotics perspective. We have decided to limit our investigation to exploration and navigation problems. In these activities, a swarm must search a given area of interest, either by covering it in all its parts or by focusing in precise points of interest, connecting them to some 'home' location. In all the studies presented below, the system has to deal with the absence of an absolute positioning system and of a common reference frame. Control is completely distributed and robots normally execute very simple individual behaviors. Coordination and cooperation among robots is generally required in order to cope with these limitations. We classify the swarm robotics studies in three main approaches: (i) coverage, (ii) chain formation, and (iii) communication assisted navigation. In the following sections, we give a generic description of the specific approach, and then we review the relevant work within it.

7.3.1 *Coverage in swarm robotics*

In swarm robotics, coverage refers to the problem of deploying a swarm of robots with the dual goal of maximizing the covered area and of keeping neighbors in communication range to keep the network fully connected. The first to define this problem has been discussed in Gage [14]. Coverage can have various scopes that we categorize in two main classes: *surveillance* and *navigation*. In both problems, a swarm of robots with limited sensing capabilities has to spread over a wide area. To compensate for the limited capabilities of individual robots, the swarm organizes in a network that fully covers the environment to detect possible changes or anomalies and keeps communication between nodes to spread information. In surveillance problems, the network itself carries out the desired task monitoring the environment; differently, in navigation, the resulting network is a support structure for aiding other agents (e.g., robots or humans) to complete their task. In this case, the resulting network is exploited by other agents to localize and to move in environments where unassisted navigation might be challenging, e.g., unknown or dynamic environments.

Several solutions to this problem have been proposed. We classify them according to the deployment method. We identify two main classes: artificial force based and incremental deployment.

Artificial force based. Methods based on artificial forces consider each robot as an embodied particle which exerts virtual forces on other robots. The motion of each individual robot is controlled by the resultant virtual force imposed by its neighboring robots and other components of the system. After a certain period, the system converges towards an equilibrium state in which the forces between robots are minimized. This virtual force is calculated on-board by each robot in a distributed and asynchronous way.

Additionally, this approach does not require models of the environment, localization, or communication between nodes. The only requirement is that the robots can perceive the relative position of the other robots in a local range. As a result, the algorithm is highly scalable. A further positive characteristic of this method is its simplicity, as the robots use a single mathematical rule to translate the sensor readings into a movement vector. As advantage, the behaviors written using this design method are robust and can be easily combined with vectorial operators. Finally, considering robots as particles subjected to external forces allows the developer to analyze the systems and to prove its properties with theoretical tools borrowed from solid scientific areas such as physics, control theory or graph theory [15].

The seminal work in this area is by Genovese et al. [16]. They propose a redeployment method for robotic sensor networks based on artificial forces of attraction and repulsion. The goal is to obtain a distribution of the robots proportional to the density of a pollutant in the environment, that is, a higher density of robots in more polluted areas. The robots are equipped with sensors that allow them to detect the pollutant concentration and the gradient direction. The robot movement is the result of the attraction towards higher pollutant densities (based on the gradient ascent) and repulsion from other robots.

Some years later, Reif and Wang [32] proposed the *social potential field.* In their work, each robot is subject to an artificial force, which is either attraction or repulsion from other robots according to the distance that separates them. The artificial force attracts towards robots further than distance d and repulses when the distance is smaller than d. The authors propose a set of heuristics to design social potential fields for achieving a variety of behaviors like clustering, covering, patrolling, etc. In the case of coverage, it is sufficient to fix the distance d as the limit of the communication range. In this way the robot formation maximizes the coverage while keeps connectivity between the robots of the swarm. A similar work of diffusion with artificial potential field force has been proposed three years later by Howard et al. [23].

Payton et al. [30] propose a coverage behavior implemented with similar attraction/repulsion forces. In this work, a part of the robots take static positions in the environment, acting as landmark to support the navigation of the other robots in the environment. Poduri and Sukhatme [31] study the coverage problem with the additional constrains of keeping a minimum number K of neighbors for each node. Their solution is based on varying the parameterization of the two forces (attraction and repulsion) in function of K. Spears et al. [34] introduce a general framework for physics-based behavior design, called *physicomimetics*. The authors also provide analytical studies of the potential energy and force balance equations. This novel artificial physics analysis provides a system parameterization technique to design networks with the desired topology. The framework is validated through a set of real robot experiments. Zavlanos and Pappas [46] provide a theoretical framework for controlling graph connectivity in mobile robot networks. By applying this framework to distributed algorithms, they study how to maintain, increase, and monitor connectivity in mobile robot networks. Vail and Veloso [39] apply a coverage method, based on artificial potential field, for RoboCup controllers. The work of Kalantar and Zimmer [25] proposes a decentralized method to the deployment of autonomous underwater vehicles. The work focuses on the coverage of specific areas of arbitrary shape, but known in advance. The proposed method is composed of various distinct phases, during which the robots interact locally to achieve the dual goal of covering the interior of the target area with an uniform distribution and of creating the formation border as similar as possible to the desired shape. The solution is based on virtual forces of attraction and repulsion, which are varied during the different phases of the process. The method is evaluated through a set of simulation experiments.

In conclusion, artificial force based methods, although simple, have shown to be suitable for a wide range of robot network applications. The systems implemented using these methods are fully decentralized, robust, scalable and require minimal sensing capabilities of the robots.

Incremental deployment. Incremental deployment algorithms position nodes one-at-a-time in an unknown environment, with each node determining its target location exploiting the previously deployed nodes. The network can either support the deployment of new nodes furnishing localization information or directly navigating them to their target positions.

Howard et al. [22] propose an incremental deployment algorithm for groups of homogeneous mobile sensors having as goal to maximize the network coverage and as constraint to maintain full connectivity. Each sensor calculates its target position by collecting all the maps generated by the already deployed nodes, merging them, and using the generated global map to calculate its position.

While the Howard's work is based on homogeneous mobile sensors, Tang et al. [38] exploit the heterogeneity of the devices: mobile sensor nodes with very limited navigational capabilities are guided to their deployment positions by more intelligent *leader* robots. Only a limited set of devices (the leaders) has capabilities to localize themselves in the (known) map, while the others are mere followers navigated to their positions. In the proposed system, as first phase, an off-line global-knowledge planner calculates the desired deployment positions of the sensors and of the leader robots, which are used as way-points during the actual deployment process. As second phase, a small team, of about 5 mobile sensors, is assigned to each leader which navigates them to preplanned positions. Iteratively, the leaders return to the base station to recruit other teams to be deployed. The navigation of the leader in the environment is facilitated by the already deployed sensors. This work has been extended by Howard et al. [24] adding to the algorithm an initial phase of exploration performed by the leader robots, which generate a global map of the environment subsequently used to plan the deployment positions. The work includes a series of experiments with real robots involving 3 leaders and 35 followers.

Stirling et al. [36] propose a fully-distributed strategy to coordinate a swarm of flying robots for indoor deployment and dynamic redeployment with particular attention on energy efficiency. Also this method is based on an incremental deployment to gradually expands the robot sensor network. Differently from the previous works, the approach is totally decentralized, does not require any reconstruction of the map, nor any exchange of large amounts of data nor any global-positioning system, but rather it relies only on relative-positioning sensors. This makes the presented algorithm scalable, with low computational complexity for any environment and swarm size.

Other methods. The following works present coverage algorithms that do not implement any of the two methods presented above. However, these works propose decentralized solutions of interesting coverage problems which are worth to be mentioned.

Winfield [43] proposes a distributed method for coverage of large environments with not sufficient resources. The robots have limited sensing and communication range, and the physically bounded environment is assumed to be sufficiently large such that cannot be entirely covered by the full connected robotic network. To overcome this constraint, the proposed solution exploits the mobility of the robots, which keep moving in the arena. The robot random motion brings them in contact, and lets the information propagate within the network.

Batalin and Sukhatme [4] compare an artificial force method based on only repulsion force with a simple rule-based local dispersion algorithm. The two investigated approaches provide similar performances.

Subsequently, the rule-based algorithm has been modified ([3]) to let mobile robots deploy small communication beacons in an unknown environment. These beacons are exploited by the robots to perform exploration of the environment.

Schwager et al. [33] present a decentralized, adaptive algorithm to deploy a network of mobile robots to an optimal sensing configuration. The sensing network has the double goal of spreading out and of maximizing the sensing metrics through aggregation in areas of high interest. The robots do not know in advance where these areas are, but they estimate this information on-line from sensor measurements. In this way, the system is decentralized and adaptive to environmental changes. Each robot combines the interest distribution estimate with the neighbors' relative position to calculate its new position. The convergence to the optimal configuration is studied through formal methods and numerical simulations. Finally, the authors show how the system achieves better results letting the robots combine their local estimate of the interest distribution with the estimations of their neighbors, resulting in a sort of collective sensing.

7.3.2 *Chain formation*

In swarm robotics, a "chain" is a linear sequence of robots in communication range with their neighbors (see Figure 7.2). Chain formation is employed to connect two locations that cannot be simultaneously perceived by the individual robots due to their limited sensing capabilities. Once formed, the chains are exploited by other robots to efficiently navigate. In this way, the robots do not need any knowledge or map of the environment, nor any absolute positioning system (e.g., GPS), but they simply follow the chain to get to desired locations.

Chain formation takes inspiration from the foraging behavior of ants. Deneubourg et al. [9] showed that ants, when foraging, deposit trails of pheromone as a form of stigmergic communication to attract other individuals. As a result, the ant colony finds the shortest path between a nest and a food source. Similarly, robotic chains support the navigation between two areas. However, instead of laying pheromone trails, robots place themselves as trail markers. Each chain begins from a predefined home location, grows in random directions and possibly degenerates, until a target location is found and a path between home and target is established. The chain formation process is completely distributed and probabilistic. Robots explore the environment through random motion and join chains at random positions when they encounter one, therefore becoming trail markers.

With a given probability, the trail marker robots at the extremity of the chain leave and possibly re-join it at a different position. In this way, the chain can grow and disband following a self-organized process that leads to a continuous exploration of the environment until the target is discovered.

The first studies of chain formation methods in artificial distributed systems have been done in simulation [8, 10, 17]. In these works, the formation of a chain supports the robots in collecting objects in an unknown environment. Subsequently, this method has been implemented also on real robots [28, 42].

Other works focus on the exploitation of already formed chains to improve the efficiency of the robotic swarm in carrying out tasks that require navigation. For example, Nouyan et al. [29] implemented on real robots a distributed system able to solve a foraging task. In the investigated scenario, the goal of the robots is to retrieve an object (*prey*) and bring it to a specific area (*nest*). To support the exploration and the navigation in the unknown environment, the swarm allocates part of the robots to form a chain that extends from the nest in search of the prey. Once found, the remaining free robots exploit the chain to get to the prey and to transport it to the nest. Figure 7.2 shows a picture of the experiment in which a robotic chain connects the nest to the prey. Another example is the work of Campo et al. [5], where a swarm robotics system exploits already formed chains to select the shortest path to a target.

Fig. 7.2 A chain of robot connecting the nest to an object to be transported.

In the work of Hauert et al. [19], a swarm of Micro Air Vehicles (MAVs) self-organizes to establish and maintain a wireless communication network between users located on the ground and the base station. The vehicles rely only on local communication and compass, without absolute nor relative position information. The controller of the MAVs has been designed through artificial evolution. The resulting behavior is a chain formation process in which the MAVs, after take off, place themselves at the beginning of the chain to extend it. They keep their position by flying in the smallest possible circular trajectory. The continuous movement of the MAVs and the intrinsic noise of sensors and actuators cause temporary disconnections from the base station. This alternation of connection and disconnection phases provides the basis for a synchronized sweeping movement of the entire swarm, which allows to explore the environment and to connect users (with unknown location) to the base station. This work is part of the *The Swarming Micro Air Vehicle Network (SMAVNET) Project*[2], during which the controller has been brought on real MAV platforms.

Ducatelle et al. [11] study the problem of redeployment, that is how to correct the positioning of the sensors in order to increase the efficiency of the whole network. In particular, they focus on the scenario of assisted navigation where ground robots receive instructions of how to move in the environment (navigation instructions) by a swarm of ceiling robots which constitutes the overhead sensor network. The ground robots have to move back and forth between two predefined locations. The general idea is that the ceiling sensors adjust their positions moving to areas effectively navigable by ground robots which have movements constrained by obstacles on the ground. The new deployment position is determined by monitoring the traffic of the robots on the ground, and moving towards directions where they travel more often. Using this simple method, eventually, the network's topology converges to a chain that connects the two locations by the shortest path. This redeployment method helps when the initial configuration of the sensors poorly matches the placement of the obstacles in the environment, which can be due either to a poor initial deployment, or to a high complexity of the scenario.

7.3.3 *Communication assisted navigation*

Similarly to chain formation, the network resulting from communication assisted navigation aims to support the navigation between two locations. In chain formation, some robots of the swarm become part of the sensor

[2]http://lis.epfl.ch/smavs

network (*sensors*), while the others exploit the network to navigate in the environment (*followers*). Differently, in communication assisted navigation, the robots that act as sensors do not take static position and do not become static landmarks, but they are sensors and receivers at the same time. While the robots move in the environment and carry out their task, they locally communicate with the other robots in order to mutually get and give navigation instructions. The integration of these instructions allows the swarm to move efficiently in the environment.

Gutiérrez et al. [18] propose the *social odometry* algorithm, which produces a self-organized collective behavior that lets groups of robots navigate more efficiently in the environment. Social odometry works as follows. Once a robot has visited a target location, it keeps up to date an estimate of its position by exploiting odometry. In addition to this information, the robot stores a confidence level, which represents how precise its estimation is. Confidence decreases as the travelled distance increases, in order to take odometric errors into account. Social odometry guides the robot to its target by combining its location estimate with the information gathered from other robots, in particular with the neighbors' estimated target locations, their respective confidence levels and the neighbors' relative positions. When the robots are numerous enough, each one is always in communication range with some other neighbors. In this way, the global result at the group level is a dynamic network between the two locations. This network aids the robots to filter odometry errors, and to improve navigation. The authors test social odometry in groups of simulated robots (varying the size up to 30 robots), and show how this technique improves the efficiency of the navigation between two locations.

Sperati et al. [35] study, through artificial evolution, strategies of efficient exploration and navigation for a swarm of robots. The task to be accomplished requires the robots to explore an unknown environment, find two distant target locations, and efficiently navigate between them. The synthesized controller resulting from the evolutionary process lets the robots self-organize in a *dynamic chain*, where the robots form two lines and keep moving between the two areas in opposite directions. Forming dynamic chains allows each individual robot to effectively travel back and forth between the two areas and allows the swarm, as a whole, to preserve the information of the area locations. Figure 7.3 shows a graphical representation of the process.

Ducatelle et al. [12] propose the *swarm navigation* algorithm in which the robots of the swarm, as in the previous work, navigate back and forth between two targets. When the robots meets with each other, they ex-

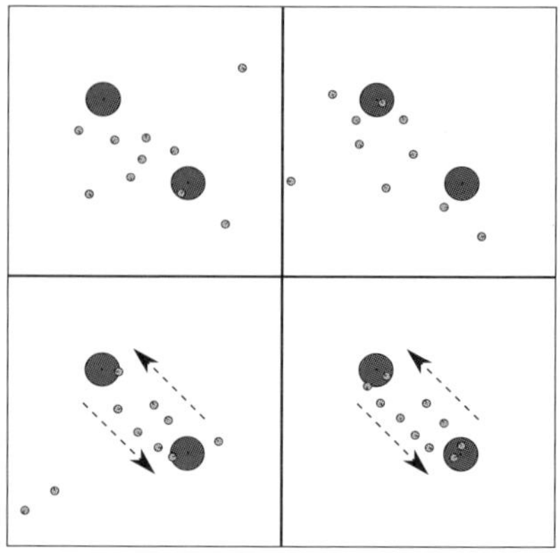

Fig. 7.3 The self-organized formation of a dynamic structure that connects two areas of interest, resulting from evolutionary optimization of the parameters of a simple neural network [35]. Robots initially start randomly distributed and progressively build a dynamic structure that connects the areas of interest.

change information of the target positions. If the robots are aiming to two different locations, they give each other navigation information about their respective targets, while if the robots aims to the same location they compare their estimate of the target position, select the newest of the two and start to move towards the same direction. This navigation strategy results in clusters of robots in communication with each other moving in opposite directions. When new information is received by any of the robots of a cluster, it spreads throughout the group, and the full cluster corrects its direction. If the number of robots is high enough, the clusters cover the entire distance between the two locations. At that point, the swarm organizes into a dynamic chain, where all the robots are in continuous movement towards one of the two locations. Keeping the robotic network in contact with the two locations lets the information of the locations position spread among all the robots of the network. In this way, the robots are continuously updated with perfect information about the direction where to move. This work has strong similarities with a previous work from Vaughan et al [40], however here the authors focus their attention only on the aspect of navigation, without any investigation on the emerging behavior of dynamic chains.

7.4 Discussions

The analysis of the current approaches in swarm robotics for exploration and navigation reveals many possible points of contact with the WSN domain. In some cases, the solutions proposed for robotic swarms have already been translated to mobile sensors, possibly with some modification, as is the case for the potential fields approach. In other cases, such as in the network-based navigation, routing paradigms and information sharing between neighboring robots are inspired from the WSN domain, and can inform novel developments with limited effort to prove convergence and stability properties, whenever that is required. Also for the chain formation approach a translation to mobile sensors is very straightforward. Despite the fact that chains do not provide a large coverage of an area, they can be exploited in mobile WSNs for tracking objects or for discovering and monitoring a limited number of points of interest. This approach is surely interesting, and similar studies are currently under development [13].

Differences between swarm robotics and WSN are mainly in the fact that swarm robotics solutions are based on fewer assumptions, which may lead to larger applicability to different domains and environmental conditions. On the other hand, in swarm robotics there is less emphasis on the demonstration of convergence properties, which are usually verified experimentally at least in a statistical sense, rather than formally demonstrated. Apart from that, it is evident that the solutions we reviewed for coverage, exploration and navigation are applicable to mobile sensors, above all when energy constraints are not tight, that is, when the hardware platform has sufficient autonomy, when batteries can be easily replaced or energy can be harvested from the environment. If energetic issues are constraining, then mobility is certainly expensive and must be limited, much as it should be limited the processing power of the single units. For this reason, among the different approaches for distributed multi-robot systems, swarm robotics is the optimal choice as it delivers simple controllers that exploit only limited sensing and processing abilities. Other approaches are much more demanding on the required processing power (e.g., multi-robot SLAM algorithms [21]), and may not be suitable for real word applications.

Finally, we believe that the best opportunities can be given by an hybridization of WSNs with robots [26]. In particular, with respect to swarm robotics the WSN may provide a useful infrastructure that can be exploited by the robots for minimal coordination and collaboration strategies. For instance, immobile sensor nodes can enhance the environment in which robotic swarms live, and can provide a mean to implement indirect coordination strategies based on stigmergy that would be otherwise difficult to implement. In this sense, WSN nodes can store structured information locally, which would facilitate the accomplishment of the task for the swarm robotic system. The advantages of such an hybridization between swarm robotics and WSNs are twofold. On the one hand, the robotic swarm can take care of deploying and redeploying tasks to optimize the network performance with respect to the events to be monitored, as well as to optimize the energy consumption across the network by replacing nodes with high consumption with less utilized ones (e.g., to mitigate the sink-hole problem), or to repair/replace malfunctioning nodes. On the other hand, the robotic swarm can exploit the WSN and the communication infrastructure to adaptively allocate tasks among robots, and to switch between exploration and exploitation in order to deliver the optimal number of robots to each task. Decision making processes can be therefore emergent form the interaction between mobile robots and immobile nodes, and can be developed following the properties of swarm robotics systems [5]. In summary, we believe that swarm robotics can greatly enhance the deployment and the redeployment of WSNs, which at the same time can offer to swarm robotics some environmental infrastructure that can be key to implement self-organized strategies for optimal group behavior.

7.5 Conclusions

In this chapter, we have addressed the problem of deployment and redeployment of sensing nodes in Wireless Sensor Networks, and we observed how mobility of sensors nodes can offer great advantages to enhance the coverage and the properties of the network. We have observed that mobility of sensor nodes naturally leads to robotic solutions, and in particular swarm robotics seems to be the best approach for large-scale distributed systems like mobile WSNs. Looking at WSN from a swarm robotics perspective reveals much more similarities than differences, and suggests a stronger integration of efforts in future research. Indeed, much advancement can be achieved in both domains by a concurrent development of solutions for swarm-like mobile WSNs. The research in this direction has just started, and much improvement is expected in the years to come, especially in the direction of delivering hybrid solutions for practical applications.

References

[1] Bartolini, N., Calamoneri, T., La Porta, T. F. and Silvestri, S. (2011a). Autonomous deployment of heterogeneous mobile sensors, *IEEE Transactions on Mobile Computing* **10**, 6, pp. 753–766.

[2] Bartolini, N., Calamoneri, T., Massini, A. and Silvestri, S. (2011b). On adaptive density deployment to mitigate the sink-hole problem in mobile sensor networks, *Mobile Networks and Applications* **16**, 1, pp. 134–145.

[3] Batalin, M. A. and Sukhatme, G. S. (2002a). Sensor Coverage using Mobile Robots and Stationary Nodes, in *SPIE on Scalability and Traffic Control in IP Networks II (Disaster Recovery Networks)* (SPIE Press), pp. 269–276.

[4] Batalin, M. A. and Sukhatme, G. S. (2002b). Spreading Out : A Local Approach to Multi-robot Coverage, in *The 6th International Symposium on Distributed Autonomous Robotics Systems (DARS)* (Springer Verlag, Berlin, Germany), pp. 373–382.

[5] Campo, A., Gutiérrez, Á., Nouyan, S., Pinciroli, C., Longchamp, V., Garnier, S. and Dorigo, M. (2010). Artificial pheromone for path selection by a foraging swarm of robots, *Biological Cybernetics* **103**, 5, pp. 339–352.

[6] Chang, C.-Y., Sheu, J.-P., Chen, Y.-C. and Chang, S.-W. (2009). An obstacle-free and power-efficient deployment algorithm for wireless sensor networks, *IEEE Transactions on Systems, Man, and Cybernetics, Part A: Systems and Humans* **39**, 4.

[7] Chellappan, S., Bai, X., Ma, B., Xuan, D. and Xu, C. (2007). Mobility Limited Flip-Based Sensor Networks Deployment, *IEEE Transactions on Parallel and Distributed Systems* **18**, 2, pp. 199–211.

[8] Cohen, W. W. (1996). Adaptive Mapping and Navigation by Teams of Simple Robots, *Robotics and Autonomous Systems* **18**, 4, pp. 411–434.

[9] Deneubourg, J.-L., Aron, S., Goss, S. and Pasteels, J.-M. (1990). The self-organizing exploratory pattern of the argentine ant, *Journal of Insect Behavior* **3**, 2, pp. 159–168.

[10] Drogoul, A. and Ferber, J. (1992). From Tom Thumb to the Dockers : Some Experiments with Foraging Robots, in *From animals to animats 2. Proceedings of the 2nd international conference on simulation of adaptive behavior (SAB92)* (MIT Press, Cambridge, MA, USA), pp. 451–459.

[11] Ducatelle, F., Di Caro, G. A., Pinciroli, C. and Gambardella, L. (2011a). Self-organised cooperation between robotic swarms, *Swarm Intelligence* **5**, 2, pp. 73–96.

[12] Ducatelle, F., Di Caro, G. A., Pinciroli, C., Mondada, F. and Gambardella, L. (2011b). Communication assisted navigation in robotic swarms: self-organization and cooperation, in *Proceedings of the IEEE/RSJ International Conference on Intelligent Robots and Systems (IROS)* (IEEE Press, Los Alamitos, CA, USA), pp. 4981–4988.

[13] Erdelj, M., Razafindralambo, T. and Simplot-Ryl, D. (2013). Covering Points of Interest with Mobile Sensors, *IEEE Transactions on Parallel and Distributed Systems* **24**, 1, pp. 32–43.

[14] Gage, D. W. (1992). Command Control for Many-Robot Systems, in *Nineteenth Annual AUVS Technical Symposium* (The Association of American Publishers, Washington, DC, USA), pp. 28–34.

[15] Gazi, V. and Passino, K. M. (2004). Stability analysis of social foraging swarms. *IEEE Transactions on Systems, Man, and Cybernetics* **34**, 1, pp. 539–57.

[16] Genovese, V., Dario, P., Magni, R. and Odetti, L. (1992). Self Organizing Behavior and Swarm Intelligence in a Pack of Mobile Miniature Robots in Search of Pollutants, in *IEEE/RSJ International Conference on Intelligent Robots and Systems (IROS)* (IEEE Press, Los Alamitos, CA, USA), pp. 1575–1582.

[17] Goss, S. and Deneubourg, J.-L. (1991). Harvesting By A Group Of Robots, in *The First European Conference On Artificial Life* (MIT Press, Cambridge, MA, USA), pp. 195–204.

[18] Gutiérrez, A., Campo, A., Santos, F. C., Monasterio-huelin, F. and Dorigo, M. (2009). Social Odometry : Imitation Based Odometry in Collective Robotics, *International Journal of Advanced Robotic Systems* **6**, 2, pp. 129–136.

[19] Hauert, S., Zufferey, J.-C. and Floreano, D. (2009). Evolved swarming without positioning information: an application in aerial communication relay, *Autonomous Robots* **26**, pp. 21–32.

[20] Heo, N. and Varshney, P. K. (2005). Energy-Efficient Deployment of Intelligent Mobile Sensor Networks, *IEEE Transactions on Systems, Man and Cybernetics, Part A: Systems and Humans* **35**, 1, pp. 78–92.

[21] Howard, A. (2006). Multi-robot simultaneous localization and mapping using particle filters, *The International Journal of Robotics Research* **25**, 12, pp. 1243–1256.

[22] Howard, A., Matarić, M. J. and Sukhatme, G. S. (2002a). An Incremental Self-Deployment Algorithm for Mobile Sensor Networks, *Autonomous Robots* **13**, 2, pp. 113–126.

[23] Howard, A., Matarić, M. J. and Sukhatme, G. S. (2002b). Mobile Sensor Network Deployment using Potential Fields: A Distributed, Scalable Solution to the Area Coverage Problem, in *The 6th International Symposium on Distributed Autonomous Robotic Systems DARS02* (Springer Verlag, Berlin, Germany), pp. 299–308.

[24] Howard, A., Parker, L. E. and Sukhatme, G. S. (2006). Experiments with a Large Heterogeneous Mobile Robot Team: Exploration, Mapping, Deployment and Detection, *International Journal of Robotics Research* **25**, 5-6, pp. 431–447.

[25] Kalantar, S. and Zimmer, U. R. (2007). Distributed shape control of homogeneous swarms of autonomous underwater vehicles, *Autonomous Robots* **22**, 1, pp. 37–53.

[26] Li, X., Falcon, R., Nayak, A. and Stojmenovic, I. (2012). Servicing wireless sensor networks by mobile robots, *IEEE Communications Magazine* **50**, 7.

[27] Ma, K., Zhang, Y. and Trappe, W. (2008). Managing the mobility of a mobile sensor network using network dynamics, *IEEE Transactions on Parallel and Distributed Systems* **19**, 1, pp. 106–120.

[28] Nouyan, S., Campo, A. and Dorigo, M. (2008). Path formation in a robot swarm, *Swarm Intelligence* **2**, 1, pp. 1–23.

[29] Nouyan, S., Gross, R., Bonani, M., Mondada, F. and Dorigo, M. (2009). Teamwork in Self-Organized Robot Colonies, *IEEE Transactions on Evolutionary Computation* **13**, 4, pp. 695–711.

[30] Payton, D. W., Daily, M., Estowski, R., Howard, M. and Lee, C. (2001). Pheromone Robotics, *Autonomous Robots* **11**, 3, pp. 319–324.

[31] Poduri, S. and Sukhatme, G. (2004). Constrained coverage for mobile sensor networks, in *IEEE International Conference on Robotics and Automation (ICRA)* (IEEE Press, Piscataway, NJ, USA), pp. 165–171.

[32] Reif, J. and Wang, H. (1999). Social Potential Fields: A distributed behavioural control for autonomous robots, *Robotics and Autonomous Systems* **27**, 3, pp. 171–194.

[33] Schwager, M., Rus, D. and Slotine, J.-J. (2009). Decentralized, Adaptive Coverage Control for Networked Robots, *The International Journal of Robotics Research* **28**, 3, pp. 357–375.

[34] Spears, W. M., Spears, D. F., Hamann, J. C. and Heil, R. (2004). Distributed, physics-based control of swarms of vehicles, *Autonomous Robots* **17**, 2-3, pp. 137–162.

[35] Sperati, V., Trianni, V. and Nolfi, S. (2011). Self-organised path formation in a swarm of robots, *Swarm Intelligence* **5**, 2, pp. 97–119.

[36] Stirling, T., Wischmann, S. and Floreano, D. (2010). Energy-efficient indoor search by swarms of simulated flying robots without global information, *Swarm Intelligence* **4**, 2, pp. 117–143.

[37] Suzuki, T., Sugizaki, R., Kawabata, K., Hada, Y. and Tobe, Y. (2010). Autonomous deployment and restoration of sensor network using mobile robots, *International Journal of Advanced Robotic Systems* **7**, 2.

[38] Tang, Y., Birch, B. and Parker, L. E. (2004). Planning Mobile Sensor Net Deployment for Navigationally-Challenged Sensor Nodes, in *IEEE International Conferences on Robotics and Automation (ICRA)* (IEEE Press, Piscataway, NJ, USA), pp. 172–179.

[39] Vail, D. and Veloso, M. (2003). Multi-Robot Dynamic Role Assignment and Coordination Through Shared Potential Fields, in *Multi-Robot Systems* (Kluwer, Hingham, MA, USA), pp. 87–98.

[40] Vaughan, R. T., Støy, K., Sukhatme, G. S. and Matarić, M. J. (2002). LOST: Localization-Space Trails for Robot Teams, *IEEE Transactions on Robotics and Automation* **18**, 5, pp. 796–812.

[41] Wang, G., Cao, G. and La Porta, T. (2006). Movement-assisted sensor deployment, *IEEE Transactions on Mobile Computing* **5**, 6, pp. 640–652.

[42] Werger, B. B. and Matarić, M. J. (1996). Robotic "Food" Chains: Externalization of State and Program for Minimal-Agent Foraging, in *From Animals to Animats 4. Proceedings of the 4th international conference on Simulation of Adaptive Behavior (SAB96)* (MIT Press, Cambridge, MA, USA), pp. 625–634.

[43] Winfield, A. F. T. (2000). Distributed Sensing and Data Collection Via Broken Ad Hoc Wireless Connected Networks of Mobile Robots, in *Distributed Autonomous Robotic Systems*, Vol. 4 (Springer Verlag, Berlin, Germany), pp. 273–282.

[44] Wu, C.-H., Lee, K.-C. and Chung, Y.-C. (2007). A Delaunay Triangulation based method for wireless sensor network deployment, *Computer Communications* **30**, 14-15, pp. 2744–2752.

[45] Yang, Y. and Cardei, M. (2007). Movement-assisted sensor redeployment scheme for network lifetime increase, in *Proceedings of the 10th ACM Symposium on Modeling, analysis, and simulation of wireless and mobile systems (MSWiM '07)* (ACM Press, New York, NY, USA).

[46] Zavlanos, M. M. and Pappas, G. J. (2007). Potential Fields for Maintaining Connectivity of Mobile Networks, *IEEE Transactions on Robotics* **23**, 4, pp. 812–816.

Chapter 8

Robot Cooperation and Swarm Intelligence

Nicole El Zoghby[1], Valeria Loscrí[2], Enrico Natalizio[1] and Véronique Cherfaoui[1]

[1] *Heudiasyc Lab., UMR CNRS 7253, Université de Technologie de Compiègne, France,*
[2]*DEIS, Universitá della Calabria, Italy*

Abstract. *This chapter is devoted to illustrate and characterize the relationship between Swarm Intelligence and cooperation among robots. Individuals with very limited computational capabilities are able to carry out very complex tasks when they can work together. From a methodological point of view, Swarm Intelligence is a set of heuristic solutions inspired by animal swarm behaviors and capable to offer empirical solutions to many computationally hard problems pertaining to several disciplines. In this chapter, we will try to outline the main research directions in Swarm Intelligence implementation within a robot network through the cooperation among the robots. The latter topic will be presented along with its advantages, issues and challenges. The convergence of robot cooperation and Swarm Intelligence is leading towards a new discipline, called Swarm Robotics. In this chapter, we will introduce this new field of study, its most relevant works and its main research directions.*

8.1 Introduction

Swarm Intelligence is a powerful concept that pivots around the cooperation among the members of a community towards a common goal. From a methodological point of view, Swarm Intelligence is a set of heuristic solutions inspired by animal swarm behaviors and capable to offer empirical solutions to many computationally hard problems pertaining to several disciplines. In this chapter, we will focus on the relationship between Swarm Intelligence and cooperation among robots. We will start presenting Swarm Intelligence, its biological principles and the mechanisms that underlie collective behaviors, the most important Swarm Intelligence heuristics and their classical applications. After this introduction, we will survey the state of the art on cooperation among robots in order to present advantages, issues and challenges of this research field. Finally, we will try to bridge swarm intelligence and cooperation among robots towards the description of a very recent discipline: Swarm Robotics, of which we will show existing taxonomies and applications.

8.2 Swarm Intelligence

The word "swarm" evokes the image of a large number of small insects where each individual performs a simple task, but whose action produces a complex behavior as a whole [39]. The emergence of such a complex behavior extends beyond the swarms. Complex social structures are similar in bigger animals as well as other types of insects. Some examples are colonies of ants and termites, flocks of birds, schools of fish, colonies of bacteria, or even herds of terrestrial animals. Swarms are defined as collections of many simple individuals that interact with both other individuals and the surrounding environment [60]. The combination of their simple or microscopic behaviors causes considerably more complex and macroscopic actions, which enable the whole system to achieve remarkable results as a whole.

The term Swarm Intelligence was introduced for the first time by Gerardo Beni and Jing Wang in 1989 [9]. The Swarm Intelligence studies the collective behavior of systems composed of many individuals who interact locally with each other and with the surrounding environment, using forms of decentralized and self-organized control to achieve their objectives.

Therefore, the Swarm Intelligence provides a new framework for the design and implementation of systems consisting of many agents that are able to cooperate in order to solve highly complex problems. The potential benefits of such approach are several:

- robustness: the failure of individual elements does not degrade significantly the performance of the entire system;
- simplicity: the individual behavior is simple but still it allows to reduce the complexity of individuals;

- scalability: the control mechanisms used do not depend on the agents number within a swarm.

8.2.1 *Biological principles for swarm intelligence*

The Swarm Intelligence comes from biological insights related to the enormous capabilities that social insects possess to solve daily-life problems within their colonies. Insects belong to two families: the most ancient is the "Isoptera" (termites) entirely social, the second is the Hymenoptera, which includes ants, wasps, bees and also presents social structures. These insects, even if they belong to two distinct families, which are very far from each other in evolution, share three important characteristics:

(1) individuals of the same species cooperate in the care for the young ones;
(2) individuals share the reproductive division of labor, sterile individuals work for the benefit of fertile individuals;
(3) an overlap of at least two generations exists, so that the offspring can help parents to carry out the tasks necessary to the life of the colony.

On the basis of these three characteristics, the entomologists distinguish the true sociability, or *eusociality*, from behaviors that do not present all the three listed characteristics, and for this reason are defined pre-social. In what follows, we will analyze the main biological principles that govern the organization in the colony of insects, that is the mechanisms which give rise to complex collective behavior of social insects, the concept of stigmergy and the theory of self-organization in biological systems. After this, we will introduce the main metaheuristics of Swarm Intelligence.

Ants, wasps and termites are able to build sophisticated nests in cooperation, even if none of the individuals have an exact plan of how to proceed and no coordinator exists [11]. Another example is taken from the behavior of ants and bees during the search for food. The ants employ a strategy of indirect communication through the release of a chemical substance, called *pheromone*, in order to identify the shortest paths between nest and food sources (Fig. 8.1);

Bees are very efficient in finding the richest sources of food by using some explorers that communicate the information about newly-found sources of food by a *waggle dance* (so-called for the vibrations generated from the abdomen of the bees while flying) (Fig. 8.2).

An African species of termites, the *Macrotermes bellicosus*, builds mounds that can reach 30 m of diameter and 6 m of height [70] (Fig. 8.3).

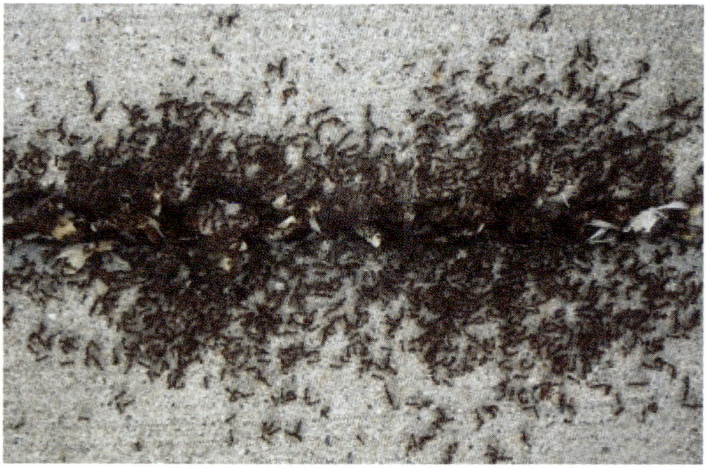

Fig. 8.1 A group of ants following a pheromone trail.

Fig. 8.2 Bees' waggle dance.

These "skyscrapers" are the result of the biological work of millions of tiny individuals (1-2 mm long), which are completely blind. Even more fascinating than the size of these mounds is their internal structure.

The nests of the species *Apicotermes lamani* are probably one of the most complex structures ever built in the animal kingdom. The hive is a highly complex structure, high around 20 to 40 cm. On the outer surface there are a series of micro-structures which provide the air-conditioning and gas exchange with the external environment, while inside the hive, rooms are concatenated to each other by means of helical ramps. These spiral ramps are born from the twisting and welding of consecutive floors. There

Fig. 8.3 Mound built by Macrotermes bellicosus.

are different stairs on each floor and some of them cross the entire nest. Hence, even the most distant rooms are connected by these shortcuts. The complexity of these structures and the collective behavior do not reflect the relative simplicity of the individual behavior of a single insect. Of course, insects are complex entities, able to adapt their behavior according to many sensor inputs. However, the complexity of a single insect in terms of cognitive or communicative skills may be high according to an absolute perception, but it is not sufficient to control a large system and explain the complexity of all the behaviors that govern a colony [73]. In essence, a single insect is not able to find an efficient solution by itself to a problem of the colony, while the group to which it belongs manages to find, as a whole, a solution very easily. Behind this organization with no boss, there are several hidden mechanisms that enable groups of insects, whose members have to deal only with partial information about their surroundings, to face random situations and find solutions to complex problems.

8.2.1.1 *Mechanisms for collective behavior*

The study of the mechanisms that underlie the collective behavior of insects started more than a century ago. Initially, in order to justify the complexity of these behaviors, it was assumed that the individual insects possessed a minimum knowledge of the overall structure that needed to be produced and that, accordingly, they were able to make the appropriate decisions. In other words, it was thought that there was a causal relationship between the complexity of decisions, the patterns observed at the level of the colony, and the behavioral and cognitive complexity that was supposed to be required at the individual level to make these decisions and models. Therefore, it was assumed that the model which governed those companies was hierarchical and centralized. However, most of the research done in recent years, has revealed a completely different organization. Today, we know that the individual insects do not require representation, schema, or explicit knowledge of global structure they produce. A single insect is not able to evaluate the global situation to centralize the information about the state of the entire colony and later to control the tasks that need to be made by other workers. There is no supervisor in these colonies. A colony of social insects is quite similar to a decentralized system composed of autonomous units that are distributed in the environment, and could be described by simple probabilistic cause-effect behaviors [70].

The principles that underlie the interactions between insects are carried out through local information of a global model. Each insect follows a set of few rules. For example, the ants can perform, on average, approximately 20 different elementary behaviors. At the colony level, the organization emerges from the interactions that occur among individuals who show these simple behaviors. These interactions ensure the propagation of information within the colony and also organize the activity of each individual. With these sophisticated networks of interactions, social insects can solve a wide range of problems and respond to external challenges in a very flexible and robust way.

8.2.1.2 *Stigmergy*

The first scientific explanation of activities organization of social insects was given 40 years ago by the French biologist Pierre-Paul Grasse, who introduced the concept of *stigmergy* to explain some of his observations on the behavior of termites in the construction of termite mounds [70]. The self-organization of social insects requires interactions among themselves. This interaction may be direct or indirect. Direct interactions are obvious: sight contact and/or chemical; whereas indirect interactions are more subtle: two individuals interact indirectly when one of them modifies the environment and the other responds accordingly to the new environment [29]. This interaction is an example of stigmergy. This term, which comes from the greek words "stigma" that means sign and "ergon" that means work (led by stimuli), is a form of indirect communication in which each individual acts on the surrounding environment and other individuals that detect some changes in the environment react to the stimulus. Since the overhead of communications does not increase when the size of the group increases, the stigmergy allows great scalability. It should be noted that the stigmergy in itself does not explain how the communication takes place indirectly, but only provides a general mechanism that relates the behavior of the individual to the level of the colony. In his studies, Grasse showed that the coordination and regulation of activities of a colony did not depend on the workers, but they were mainly driven by the nest. In other words, the information coming from the local environment and the progress of the work can drive the individual activity. Each time that a worker performs an action, this action results in a modification of the local configuration. The new configuration will affect other subsequent actions and behaviors of the other workers in the colony. This process leads to an almost perfect coordination of collective labor and may give us the impression that the colony is following a definite plan.

A good example of stigmergic behavior is the search for food of the ants. The ants communicate with each other through the use of pheromones, chemicals that attract other ants. When an ant finds a source of food, it quickly returns to the nest and releases a pheromone trail. This trail will then lead the others from the nest to the food source. While returning to the nest, the ants release their pheromones along the path, thus reinforcing the trail. The formation of the trail therefore derives from a positive feedback: the greater the number of ants that follow the path, the more the path will become attractive and appealing. Of course, the trail will disappear after a while, if the reinforcement is too weak, and this may happen when the food source is exhausted. The interesting thing is that this system of maintenance of the trail is not only a mechanism used to quickly gather a large number of purveyors around the source of food, but it also allows the colony to take efficient decisions such as the choice of the shortest path that leads to the source of food. From this description, other properties characterizing the stigmergy have emerged. In fact, stigmergy affects the overall behavior of the population by two key elements of self-organization that have already been implicitly introduced: the positive feedback and its dual, the negative feedback. The positive feedback is the phenomenon by which the marks on the environment, deployed by individuals, encourage other members to release additional marks in the same place, making the population converge toward the reinforcement of the solution. The negative feedback is the opposite phenomenon, i.e., areas marked weakly tend to be overlooked by the individuals of the population, leading to the impoverishment of the solution. Another property is that the emerged stigmergy is a form of communication limited in time, i.e., the changes on the environment vanish over some time. For example, the pheromone released by ants evaporates with time, therefore, it is necessary to operate the reinforcement of the path to keep it alive.

8.2.1.3 *Principles of self-organization*

Self-organization is a set of dynamic mechanisms by means of which the structure of a system appears at the global level as a set of interactions of its components at the local level. It has four basic components: the positive feedback that derives from the execution of simple behavioral rules that support the creation of structures; the negative feedback that counteracts the positive feedback and leads to the stabilization of the collective behavior; the amplification of fluctuations through the positive feedback; direct and multiple interactions or stigmergic interactions among individuals to produce deterministic results and the appearance of large-scale durable structures. In addition to the components described so far, the self-organization is also characterized by some key properties:

(1) Self-organizing systems are dynamic. As mentioned previously, the production of structures, as well as their persistence, requires constant interactions among the permanent members of the colony and their surroundings.
(2) Self-organizing systems exhibit emerging properties. They show more complex properties of the single contribution of each individual. These properties arise from a combination of non-linear interactions between the members of the colony.
(3) Together with emergent properties, nonlinear interactions lead self-organized systems to bifurcations. A bifurcation is the appearance of new stable solutions when there is a change of the parameters of the system. This corresponds to a qualitative change in the collective behavior.
(4) Finally, the self-organizing systems can be multi-stable. Multi-stability means that, for a given set of parameters, the system is able to achieve different stable states that depend on the initial conditions and random fluctuations.

8.2.1.4 *Collective behaviors*

The processes of self-organization described above may produce a wide variety of collective behaviors that are intended for the resolution of a given problem. In their studies, Camazine et al. [15] have proposed to categorize social behaviors of a colony of insects, according to four types of tasks: individual, group, team and shared tasks. Following this categorization, each global task in the colony (e.g., nest building) can be hierarchically split into sub-tasks belonging to any of the mentioned types. This method can be seen as the decomposition of a problem into simpler tasks, which are essential for the resolution of the problem. Another way to characterize social insects collective behavior consists in defining specific functions to describe insects' tasks. It is possible to identify four main categories of functions: coordination, cooperation, evaluation and collaboration. These categories are not mutually exclusive and they contribute together to the fulfillment of the various collective tasks of the colony. Below, we provide a first definition of each of these functions and subsequently explain their respective roles in some examples of collective behavior of social insects.

Coordination. Coordination is the appropriate organization in space and in time of all the tasks necessary to solve a specific problem. This leads to specific spatio-temporal distributions of individuals, of their activities and/or outcomes of their activities, in order to achieve a certain goal. For example, the coordination occurs in the organization of the movement of swarms of bees and locusts. In this case, the interactions between individuals generate synchronized movements (temporal organization) and oriented (spatial organization) of individuals towards a specific goal.

Cooperation. The cooperation is a phenomenon that occurs when a task can not be performed by a single individual but requires a set of them. Therefore, individuals must combine their efforts in order to successfully solve a problem that goes far beyond their individual capabilities. For example, cooperation is required from ants to remove a long wooden stick that obstructs the entrance of their nest. In this situation, the ants combine their efforts to pull the stick away from the hole. Some ants raise the stick while others put their head inside the entrance in order to avoid that the stick can fall back inside. In the end, the combined efforts lead the group to remove the stick. The function of cooperation represents the mechanisms that go beyond the limitations of individuals.

Evaluation. The term evaluation refers to the mechanisms that occur when a colony is faced with several opportunities. These mechanisms are the result of a collective choice of at least one of opportunity. For example, when the ants Lasius niger find different food sources, or different routes that lead to one food source, they generally choose only one among the various possibilities. The evaluation is usually guided by competition between the chemical trails that underlie each possibility. In most cases, the ants will end up choosing the richest food source reach via the shortest path.

Collaboration. With the term collaboration we mean the various activities which are carried out simultaneously by a group of specialized individuals. This specialization stems from both a behavioral and morphological differentiation of individuals. The most striking expression of this division of labor is the existence of castes. For example, the workers in a colony of ants that shred the leaves may belong to four different castes and their size is closely related to the tasks they perform. Only the workers who have a head size greater than 1.6 mm are able to shred the leaves, which are used to grow fungi that are the primary source of food for these colonies. Conversely, only the worker ants that have a smaller head size of about 0.5 mm are able to take care of fungi cultivation.

Most of the organization of the collective behavior of social insects can be seen as the combination of the four functions of coordination, cooperation, evaluation and collaboration. Each of these functions emerge at the collective level by the continuous interactions among insects. Together, the four functions of organization produce solutions to the problems of the colony and could give the impression that the colony act as a whole, planning the work to achieve the colony's goals.

8.2.2 *Main meta-heuristics of swarm intelligence*

The main disadvantage present in the algorithms that rely on constructive methods or iterative improvements is that they generate only a limited number of solutions [26]. In order to try to solve these problems, it is possible to use metaheuristics. The term metaheuristic comes from two Greek words: heuristic derives from the verb heuriskein, i.e., search, while the suffix meta means beyond to a higher level [11]. A metaheuristic is a heuristic method, i.e., a general algorithm or a set of algorithmic concepts applicable to a diverse number of optimization problems that, with slight modifications, can be adapted to describe a specific problem. Below we will explain the most popular Swarm intelligence metaheuristics.

8.2.2.1 *Ant colony optimization*

Ant colony optimization (ACO) is a metaheuristic within which a colony of artificial ants cooperate in order to obtain good solutions to difficult discrete optimization problems. Cooperation is the key component of ACO algorithms. In fact these algorithms allocate computational resources to a number of relatively simple agents (artificial ants) that communicate indirectly through the stigmergy [26]. ACO algorithms can be used to solve combinatorial optimization problems, both static and dynamic. The static problems are those in which the characteristics of the problem are known a priori, when the problem is defined, and do not change during the resolution of the problem. A classic example is the TSP (traveling salesman problem), where cities and the distances among cities are part of the definition of problem and do not change during the execution time of the algorithm. The dynamic problems are defined as functions of a number of quantities, the values of which are set by the dynamics of the problem. The instance of the problem therefore changes at the execution time and, therefore, the optimization algorithm must be able to adapt to a changing environment. An example of a dynamic problem can be identified in the routing problems in networks, in which the data traffic and network topology may change very frequently.

Description.
An artificial ant is a constructive and stochastic process that incrementally builds a solution by adding appropriate components to the partial solution. Therefore, the ACO metaheuristic can be applied to some combinatorial optimization problems, for this reason, it can be defined as a constructive heuristic. If we consider an optimization problem (S, f, Ω), where S is a set of possible solutions, f is the objective function that assigns a cost to each candidate solution s belonging to the set S, and $\Omega(t)$ is a set of conditions to satisfy. The parameter t indicates that the objective function and the conditions imposed are both functions of time. The goal is to find a globally optimal solution s^*, that is a possible solution at minimal cost. From this, the ants develop artificial solutions of performing random paths on a fully connected graph $G_c = (C, L)$, called construction graph with L connections. In many applications, the ants construct feasible solutions, even if, in some cases, it is necessary to allow them to construct improbable solutions. For example, a component $c_i \in C$ and a connection $l_i \in L$ may be associated with a trace of pheromone τ (τ_i if the trace is associated only to a component, τ_{ij} if it is associated also to a connection), and a heuristic value μ (μ_i and μ_{ij}, respectively). The pheromone trace encodes a long-term memory of the entire process, and it is updated by the ants themselves. Instead, the heuristic value, often called heuristic information, is an input given *a priori* information and based on the current problem instance, or otherwise it is a run-time input provided by the ants through different sources. In many cases, μ represents a cost, or at least an estimate of the cost of adding

components or connections to the solution under certain conditions. The heuristic values are used into ants in order to make decisions on how to navigate the probabilistic graph. It is important to note that the ants act simultaneously and independently of each other. Even if it is very complex for each single ant to find a solution to the proposed problem, good quality solutions can still be obtained as a result of a collective cooperation among the ants. All this is obtained through an indirect communication represented by the traces of pheromone. In more specific terms it can be said that an ACO algorithm can be represented through a combination of three different procedures: *Construct Ants Solutions*, *Update Pheromones*, and *Daemon Actions*.

Construct Ants Solutions manages a colony of ants concurrently and asynchronously in the search for adjacent states of a given problem, moving through neighboring nodes of the problem graph G_c. They move by applying a stochastic and local policy decision, which is based on the use of pheromone trace and heuristic information. In this way, the ants are able to search for the solution to the optimization problem in an incremental manner. As soon as an ant has found a solution, or while the solution is to be added, the ant evaluates the (partial) solution that will be used during the following procedure *Update Pheromones* to determine how much pheromone must be deposited.

Update Pheromones is the process by which the traces of pheromone are updated. The importance of the path can both increase, by depositing pheromone on the components and the connections use, or decrease due to the evaporation of the pheromone itself. From the practical point of view, the deposit of new pheromone increases the likelihood that these components/connections are used again by ants future. Otherwise, the evaporation of the pheromone realizes a form of forgetfulness, in order to avoid a too rapid convergence of the algorithm towards a sub-optimal region. Therefore, it promotes the exploration of a new area within the search space.

Finally there is *Daemon Actions*, a procedure used to implement centralized actions that cannot be performed by the individual ants. Examples of such procedure are the activation of a local optimization procedure, or the collection of global information that can be used to decide whether it is useful or not to deposit additional pheromone, in order to influence the research process considering a non-local perspective. As a practical example, the daemon can observe the path found by each ant in the colony and select one or more ants, which can thereafter afford to deposit of additional pheromone on components or connections they have been using.

Applications.
ACO metaheuristic finds application in most of the cases located problems in combinatorial optimization. The most classical problem of this type is the the Traveling Salesman Problem (TSP) [59]. It is a problem of a salesman that, starting from her home city, is looking for the possible shortest paths within a given set of customer cities, visiting each city at least once before returning home. The TSP can be represented by a weighted and complete graph. TSP is a problem based on finding the shortest Hamiltonian path length in the graph, where the Hamiltonian path is a closed path that passes through each node of the graph exactly once. The trace of pheromone τ_{ij} in the TSP refers to the desire to visit node j directly after passing through the node i. The heuristic information μ_{ij} is typically inversely proportional to the distance between two cities i and j, $\mu_{ij} = 1/d_{ij}$. Each ant is initially placed on a random city and at each step it adds in its path a city that was not yet visited, The construction of the solution ends when all cities have been visited. Therefore, TSP is a NP-hard combinatorial optimization problem that attracted a lot of research. The TSP has a central role in ACO problems, in fact it has been used for Ant System, the first ACO algorithm.

8.2.2.2 *Particle Swarm Optimization*

The Particle Swarm Optimization (PSO) algorithm is defined as a population-based algorithm characterized by a set of candidate solutions, where each solution is "a particle" moving in a search space [37]. The PSO was introduced for the first time in 1995 by Kennedy and Eberhart, whose objective was to introduce a new methodology in the computational intelligence [46]. This technique uses simple analogies with social interactions, rather than purely individual cognitive skills [67]. The roots of this metaheuristic bring back to ideas and techniques investigated for computer graphics and social psychology. In the field of computer graphics, the first work that proposed a PSO algorithm can be traced in the work of Reeves (1983), who proposed particle systems to model dynamic objects that could not be easily represented by polygons and surfaces (fire, smoke, water and clouds). The social psychology, in particular the theory of dynamic social impact, was another source of inspiration for the development of the first PSO algorithm. The principle that governs the movement of a particle in a search space of a problem can also be compared with a model of human social behavior, in which individuals adapt their behaviors to satisfy those of their peers. In PSO, simple entities, called particles, are located in an area of research of a specific problem or function, and evaluate a fitness function in their current location. This fitness function is based on a more important and generic function that characterizes the behavior of the entire swarm, called the objective function. Each particle can determine the displacement within the search space by combining some aspects of the its best position history with those of one or more members of the swarm through some random perturbations. Next iteration takes place when there is an update of all the particles. Each individual particle in the swarm is characterized by three-dimensional vectors D, where D represents the size of the search space and it is associated with the following values:

- current position, x_i,
- best past position, p_i,
- particle speed, v_i.

The current position x_i can be seen as a point in the space. At each iteration of the algorithm, the current position is considered as a solution to the problem. If the location is the best obtained up to that point, its coordinates are loaded into the vector p_i. The value of the best result can be loaded in a variable named p_{best_i} in order to perform a comparison with the results that will be obtained from subsequent iterations. The goal is to preserve the best position achieved and load it in In the PSO, the single particle itself has no power to resolve the problem: the progress occurs only when a particle interacts with the other. In the process of particle swarm optimization, the velocity of each particle is updated iteratively in such a way that the particles oscillate stochastically around the value of p_i. PSO attracted a lot of attention and several different versions have been presented over the years, interested readers can refer to [30] for a survey.

Population dynamics: Algorithm

(1) Initialize an array of particles with random positions and velocities on a D-dimensional area of research.
(2) Loop
(3) For each particle, evaluate the fitness function.
(4) Compare the fitness function of the particle with its value p_{best_i}, if the current value is better than the value in variable p_{best_i}, assign the current value to p_{best_i} and put p_i equal to the current position x_i.
(5) Individuate the particle in the population that has obtained the best

fitness function and assign its position to a variable p_g.

(6) Change the speed and the position of the particle in accordance with the following equations:

$$\vec{v_i} \leftarrow \vec{v_i} + \vec{U}(0,\phi_1) \otimes (\vec{p_i} - \vec{x_i}) + \vec{U}(0,\phi_2) \otimes (\vec{p_g} - \vec{x_i}) \tag{8.1}$$

$$\vec{x_i} \leftarrow \vec{x_i} + \vec{v_i} \tag{8.2}$$

where $\vec{U}(0,\phi_i)$ is a random matrix of numbers uniformly distributed between $[0,\phi_i]$, which is randomly generated at each iteration and for each particle.

(7) If the criterion is met (usually a threshold value of the fitness function or a maximum number of iterations), exit the loop.

Population dynamics: Parameters definition

A small, but not insignificant, advantage of PSO is given by the relatively small number of parameters to be set. A fundamental parameter is the population size, this parameter is set in an empirical way on the basis of the size and perceived difficulty of the problem. The parameters ϕ_1 and ϕ_2 determine the relative magnitudes of random forces in the direction of the best particle $\vec{\phi_i}$ and of the best neighborhood $\vec{\phi_g}$, and are often called acceleration coefficients. The behavior of the PSO can change with the values of ϕ_1 and ϕ_2. Interestingly, the components $\vec{U}(0,\phi_1) \otimes (\vec{p_i} - \vec{x_i})$ and $\vec{U}(0,\phi_2) \otimes (\vec{p_g} - \vec{x_i})$ can be interpreted as attractive forces. When we change ϕ_1 and ϕ_2, we can get a PSO more reactive but possibly unstable, in which the velocity of the particles increases without any control.

Applications

The first practical application of the PSO was in the field of neural networks, which was presented with the same algorithm. Many other application areas have been explored since then, including telecommunications, control, data mining, design, combinatorial optimization, signal processing, and many others. Although the PSO has been used mainly to troubleshoot problems with a single goal and without constraints, PSO algorithms have been developed to solve problems with constraints, multi-objective optimization problems, problems with dynamic changes of the landscape, and multiple solutions. For a survey on PSO applications, interested readers can refer to [3].

8.2.2.3 *Stochastic diffusion search*

The Stochastic Diffusion Search (SDS), which was introduced by Bishop in 1989, is a research technique that uses a diffusion process to find the best fit of a given model within an area of research [10]. It is a well-characterized, robust, and global metaheuristic of the family of Swarm Intelligence, able to efficiently solve problems of research and optimization through composite structures. The SDS uses a form of direct communication between the agents, in a manner more akin to the *tandem calling* mechanism used by a particular ant species called Leptothorax acervorum. SDS is an algorithm for matching of a model based on a population of agents. Each agent takes care of the information related to the area of research in order to identify the best solution for a given model objective. The research space and the target model require to be split into micro features through a predefined set or alphabetical order [58].

Description of the algorithm SDS In general, the SDS can be easily applied to optimization problems, in which the objective function is decomposable into elements that can be assessed independently. To locate the optimal solutions for a given function objective, SDS employs a set of n agents, each of which stores a hypothesis x_i, in the range of optimal solutions. An iteration of the SDS algorithm involves testing and spread until one of the agents of the swarm does not converge to an excellent hypothesis. The agents in the SDS cooperate in a synchronous manner and appear to be subject to the steps explained below:

Algorithm 8.1 Standard SDS algorithm

```
Initialization;
repeat
  Test;
  Diffusion;
until (Termination criterium)
```

The first step sets the initial hypothesis of each agent. Generally, its value is selected randomly and uniformly within the search space. However, any information about the probable solutions available a priori, can be useful in the setting of assumptions. Then, each agent randomly selects a function f_i, with $i \in 1, \ldots, n$, and performs the evaluation of its own hypothesis $s_h \in S$. Based on this evaluation, the agents are divided into two groups: active and inactive. For the active agents we have that $f_i(s_h) = 0$, while for the inactive ones we have that $f_i(s_h) = 1$. It should be noted that, since f_i is a probabilistic function, it is possible that different evaluations of $f_i(s_h)$ give different results [43]. In the diffusion step each inactive agent randomly selects another agent to communicate with. If the latter is active, it can duplicate the hypothesis of the passive agent, hence the dissemination of information. If the selected agent is inactive, there can be no transfer of information between the two agents, and so the chosen agent will adopt a new random value. In contrast, the active agents in the standard SDS can not initiate any communication with other agents. While the iterations proceed, groups of agents assuming the same hypothesis constitute, for convergence, the largest group of agents that defines the optimal solution. The convergence is defined by two termination criteria:

- *Strong Halting Criterion:* after having determined that a group of agents larger than a threshold, verify that the size of the group is (stochastically) stable over a certain number of iterations.
- *Weak Halting Criterion:* it simply checks the stability and the minimum size of the total number of active agents (the total activity is strongly dependent on the current best solution found).

Since the tests occur with high frequency in the points of the solution space that show a good objective value, on average the agents spend more time on these optimal solutions, and at the same time, attract other agents. However, limited resources (a finite size of the population) ensure that only the best solution discovered up to that moment is able to maintain a stable group of agents. This different resources allocation allows the largest group of agents to determine the optimal solution, without requiring the individual agents to evaluate the objective function in an explicit way. Three recruitment strategies have been introduced for SDS: the passive recruitment (the standard mechanism), the active recruitment and the double recruitment. The passive recruitment has been briefly introduced above. The active recruitment is modeled on the behavior of the species of insects that swarm. They actively try to recruit other members to direct them towards a preferred direction, which can be a source of food or the selection of the site to build the nest. A practical example is the waggle dance of the bees in the hive. The waggle dance is performed to indicate to other bees the location of a promising source of food. During the deployment phase, in the active recruitment, the active agents seek the passive agents to communicate their hypotheses. Each active agent randomly contacts another agent B, if B is passive, it will be recruited from A (the hypothesis of A is communicated to B). Unlike the passive recruitment, in which the larger group, in theory, may increase disproportionately at each iteration, the active recruitment allows, at most, to double, in the size of the group, at each iteration (if any active agent chooses a passive agent). Finally, in the double recruiting, the mechanism of active and passive recruitment act simultaneously. Thus, both active and passive agents choose the agents and the hypotheses are transferred from the active agents to the passive ones. This mix of recruitment mechanisms in a system is considered to be the most truthful, from a biological point of view, and is therefore of particular interest. This may lead to a conflict in the allocation of hypotheses, so it is necessary to define some priorities among an active agent that assigns a hypothesis to a passive agent (active priority) and a passive agent that copies hypothesis from an active agent (passive priority). The SDS has a greedy assignment process, that is, once a good solution has been found, a large portion of the swarm is allocated for the operations, making these agents not available for further explorations. A mechanism that frees some of these resources without significantly affecting the stability properties of the groups of agents could increase the efficiency of SDS for many types of problems, in particular in dynamic optimization.

Applications

The SDS has been applied to many different problems of research and optimization such as the site selection for wireless networks, the identification of the sequence in bio-informatics, self-localization of mobile robots, object recognition, motion tracking of eyes and lips, and text search.

8.3 Robot Cooperation

Robot cooperation is a challenging domain that researchers have been investigating since the 1980's. It is the ability of solving a task by a group of robots. Robots cooperate as a team in order to achieve a common goal. Multi-robot cooperation increases efficiency of robots and allows the achievement of complex tasks, which cannot be accomplished by a single robot. Multi-robot cooperation comes into applications and extends research on single robot for many reasons:

- Tasks are basically too complex for a single robot to achieve because single robot is spatially limited;
- Using multiple simple robots may be cheaper and simpler than handling one complex robot;
- Multi-robot systems are more flexible and fault-tolerant than single robots acting alone.

Robots act through a cooperative behavior. They are aware of their teammates, they share goals and their actions are useful for the whole group. Cao et al. [82] have defined robots cooperative behavior over a multi-robot system as it follows: *Given some task specified by a designer, a multiple-robot system displays cooperative behavior if, due to some underlying mechanism (i.e., the "mechanism of cooperation"), there is an increase of the overall system utility.*

In other works, scientists classified multi-robot cooperative systems into two categories: active and passive. In active system, robots communicate between them in order to exchange information. They can organize their tasks and make decisions. In passive systems, there is no communications link between robots which makes the system easy to design and robust. Robots do not share information and do not make decisions together.

Robots community has been interested in this domain in the last decades, where many researchers wrote a number of surveys on multi-robot systems. Cao et al. [82] summarized the research into a taxonomy of cooperative systems. They surveyed five main research directions: group architecture, resource conflicts, origins of cooperation, learning problems, and geometric problems. L.E. Parker organized current research works in multi-robot systems by principal topic areas [64], and focused on the interaction of multiple mobile robots in chapter "Multiple mobile robots systems" of the handbook of robotics [75]. Verret gave a brief history of robotics and detailed some inspirations and influences in multi-robot systems [79]. Cai et al. explored few research fields in the multi-robot systems [14]. In this section, we will explore several robot cooperation aspects such as distributed fusion, cooperative localization and architectures and we will discuss communication effects on a multi-robot system. We will focus on active cooperative systems and consider the basic functionalities of such systems. Another aspect that should be developed is related to artificial intelligence such as control, planning and task allocation. Issues related to distributed artificial intelligence will be explored in the second part of this chapter.

8.3.1 *Communication*

Accomplishing a cooperative task needs some form of communication. Researchers distinguished between implicit and explicit communications. Implicit communication allows robots to communicate through their environment. In this case, communication is based upon the environment change

or behavior of other robots. Robots are equipped with sensors to observe the changes. Whereas, in explicit communication, robots exchange messages to transfer various information like positions, current status, future actions, etc. They devise also effective cooperative schemes.
Several works explored communication effect on performance of multi-robot systems in different cooperative tasks. They concluded that communication between robots can multiply their capabilities and can improve their efficiency. It can provide benefit for many tasks. Exchanging a small amount of information can lead to better performances.
Using wireless communication among multi-robot systems has become an important area of research [65] and a requirement for different scenarios ([20], [52]). It has contributed to the cooperative systems using explicit communication. Robots use this technology for exploration, distributed sensing or tracking, environmental monitoring and surveillance [40]. They should be able to send and receive information at any time. Even though there is no clear conclusion on which type of communication is better for robot cooperation: implicit communication can fulfill some tasks, while explicit communication can improve flexibility of multi-robot systems. Recent work took advantage of implicit and explicit communications in order to improve cooperation and competition between robots [81].

8.3.2 *Research fields*

8.3.2.1 *Distributed sensor data fusion*

Robots may be equipped with different sensors such as vision sensors (camera), sensitive sensors, distance measurement (radar and laser scanner) or position sensors (odometry, GPS, etc.). These sensors help robots discover their environment and are considered as important input for the perception task. This process consists in fusing data collected by the sensors, exploiting redundant information and reducing uncertainty. Different fusion algorithms have been developed and used in literature for single robots. Main fusion methods include weighted average method, Bayesian inference, Dempster Shafer theory, Kalman filter, fuzzy logic and neural networks. Zhao et al. presented a survey on robot multi-sensor fusion technology and explored different applications of multi-sensor fusion [85].

Multi-sensor fusion improves robots sensing and decision making while accomplishing different tasks. That is why these methods are exploited for multi-robot systems. Thanks to communication capabilities, robots can exchange information that can be used to enhance their perception performances. They exchange data with their teammates through two different approaches. In the first approach, the robot sends its own perception, in this case the fusion algorithm considers data sent by robots as it comes from an off-board sensor. In the second approach, robot sends data resulting from the fusion of all the received information. The same information can be received many times; this is what is called "data incest" [61]. Due to the latter, robots suffer from cycles of data dissemination where the same information is provided by independent sources and can be combined many times. Distributed multi-sensor fusion process should manage data incest issue and take account of the latency of distributed data and the different references systems. The main fusion method that had been appropriated to robot cooperation and that can manage all these latter are the Bayesian inference, Dempster Shafer theory and Kalman filter.

Bayesian inference combines multi-sensor information according to rules of probability theory based on observational evidences. It depends on the prior knowledge. In general, robots share their fused data. To avoid the data incest the method of covariance intersection is used ([41], [17]). For example, Santos et al. [72] developed a multi-robot cooperative object localization based on a decentralized Bayesian approach. Their method is composed of a local filter and a team filter. The local filter receives a reduced dimension representation of its teammates sample belief about the object location. The team filter receives Gaussian Mixture Model (GMM) representations of the object in the world frame, from the sensor teammates, and fuses them all performing Covariance Intersection among GMM components.

Dempster Shafer theory, which generalizes the Bayesian inference, deals with incomplete and uncertain data. It represents the knowledge by mass functions, updates the beliefs and combines the evidences. Wang Shuo et al. were interested in map-building task for multi-robot system [80]. A robot detects its environment using its own sensors and can exchange its sensing information with other robots to build a global map. Information is fused by using the Dempster operator adapted to the combination of independent sources. Authors explored the cooperative strategies in order to avoid invalid sensing information. Nowadays, different studies were interested in using Dempster–Shafer theory for distributed data fusion in vehicle networks [28]. This method is based on the cautious rule of combination [22] that allows combination of dependent sources due to its idempotent properties [86]. It can be suited for data combination in multi-robot systems.

Kalman filter uses statistical characteristics of the system model to recursively determine estimates for data fusion. It deals with dynamic models and fuses the low level redundant data of independent sources. Fused data is local to each robot and is not a result of combination. The communication regarding the robots' positions increases certainty and reduces imprecision about the robots' own poses [4]. Kalman filter was applied to improve the ball position estimation for a robotic soccer team [76] and its extensions were explored for cooperative behavior of mobile robots [68]. It is also applied for robot localization, we will detail this in the next section.

Distributed multi-sensor fusion is inevitable in different applications of multi-robot systems. Robots can receive different kinds of information from their teammates when cooperating to achieve a task. This information should be combined with the best fusion algorithm that can avoid data incest, latency and exploit the redundancy and complementarity of information.

8.3.2.2 *Cooperative localization*

Localization is an essential problem in multi-robot systems. Robots should be able to estimate their positions in order to navigate autonomously in an environment, which may be known or unknown for robots. Usually, most robots tasks require information about their positions and orientations. Distributed tasks require information about the whole group of robots or at least those detected in the scene. To supply multi-robot systems with a solution for these needs, cooperative localization has been introduced. It consists in locating each robot in a group within the same environment. Nowadays, we find localization methods based on proprioceptive sensors, inertial unit and/or GPS. Other methods take the assumption of known environment and locate robots with exteroceptive sensors.

For a strong cooperation, each robot should know positions of other members of the team. In general, they use their sensors in order to detect other robots and to recognize their environment. They communicate to exchange their pose, their maps, and the state of team. Therefore, communication capabilities allow cooperative localization. Some recent studies investigated the cooperative localization approach especially to improve localization accuracy. In the following, we will explore different examples.

Different localization methods dedicated for single robots were extended to multi-robot systems. In [32] collaborative robot localization of indoor robots is developed based on Markov Localization. Robots can localize themselves in the same environment, maps are supposed to be known. Whenever one robot detects another, probabilistic methods are used to synchronize each robot's belief. The Kalman filter and its extensions were deployed for cooperative localization in different studies. In [69], each robot shares the information regarding its own motion with the rest of the team. The Kalman filter is used to process the available positioning information from all the team members. It estimates a pose for each robot. The authors showed that the Kalman filter estimator can be distributed in a number of smaller communicating filters, one filter for each robot. Each filter processes sensor data collected by its robot and communication with other filter when two robots detect each other and measure their relative pose. Martinelli et al. [55] extended what have been developed in [69] and introduced an EKF approach by considering the most general relative observation between two robots. Karam et al. described a cooperative approach for collective localization of a heterogeneous group of vehicles where each vehicle updates its group state with its own sensor data [45]. The vehicles exchange their information about the positions of the rest of the group, then fuse it in order to obtain the global state of the system. In [50], Lee et al. presented a cooperative localization method for a multi-robot system. They incorporated different sensors such as GPS, odometer and gyro sensor to localize absolute and relative position. They utilized correlation between GPS errors and differential position data between the robots to refine their position data. In [66], the authors treated the problem of absolute localization of a team of robots for unknown initial robot positions. They proposed a particle clustering method which reduces the complexity of the overall localization algorithm.

Other methods were based on exteroceptive sensors and used communication to exchange information. Franchi et al. in [33] took the assumption that each robot is equipped with a sensor that measure the relative pose of nearby robots without their identity. They proposed a two stage localization system where data is processed by an associator, and EKF is used to isolate and treat the best estimates. The localization approach presented in [77] is based on stereo vision system helping each robot to recognize the others and its environment. The authors used serial and parallel fusion. The first approach identifies the position uncertainty of an observed robot while the second method reduces errors of the position.

Different studies explored robot formation for localization. Hidaka et al. proposed a method for optimizing the geometry of robots formation [38]. They evaluated the trace of the steady-state covariance of the robots position estimates. This method was applied on heterogeneous robots teams. The authors studied the effects of optimal formation on robot's localization.

As presented above, researchers have been interested in cooperative localization. Some of them used method initially designed for single robots. Others developed new approaches for multi-robot system. They profit from the advantages of collecting and integrating sensors information from different robots. They proved that cooperative localization can give a system better localization performance and improve robustness of localization for each robot in the group.

8.3.2.3 *Control architectures*

Control architecture is needed to handle robot control system complexity. This is because robots need to interact in a certain environment. The control architecture influences the system robustness. It should allow robots to act in real time and to control sensors and actuators. L.E. Parker described in the chapter "Multiple Mobile Robot Systems" in [75] four types of architectures: centralized, hierarchical, decentralized and hybrid.

- In **Centralized architecture** one agent controls the remaining agents in the system. The centralized controller provides a strategy for cooperation and decision making. Each robot in the team takes the commands.
- **Hierarchical architecture** is based on an approach where one robot supervises the action of a group of robots. Each robot of this group supervises another group of other robots and so on. Each robot receives a part of the task to execute.
- **Decentralized architecture** does not employ a central agent that controls all remaining agents. Robots take actions based on information detected in their environment. This architecture is very robust to failure, flexible and scalable, since control is not centralized. Each robot is responsible of its own actions.
- **Hybrid architecture** profits from the advantages of the above control strategies. In this strategy, agents are decentralized while a centralized planner supervises the team.

Different control architectures have been developed over the years. The interested readers can refer to [84] for a survey on the topic.

8.3.3 *Applications*

Cooperative robotics may be applied to different domains such as underwater and space exploration, building surveillance, large objects transportation and hazardous environment (Landmine detection, de-mining, etc.). They may also be implemented for air and underwater pollution monitoring, forest fire detection, service robotics in both public and private domains, transportation systems, search and rescue operations after large-scale disasters and the entertainment field:

- **Exploration** is an important application because it concerns areas where humans cannot access easily. Belbachir et al. treated the problem of underwater exploration where autonomous underwater vehicles exchange their information and cooperate to optimize their motions [8]. Leitner studied space applications [51], while Bautin et al. presented a frontier-based exploration method for multi-robot system [6]. In [48] air and ground robots were deployed.
- **Object transportation and manipulation** is a task where cooperation between a group of robots is essential. Robots surround the object and transport it to the desired destination by pushing it (Figure 8.4). Different works in literature treated this application and proposed different control approaches and motion planning ([83], [56]).

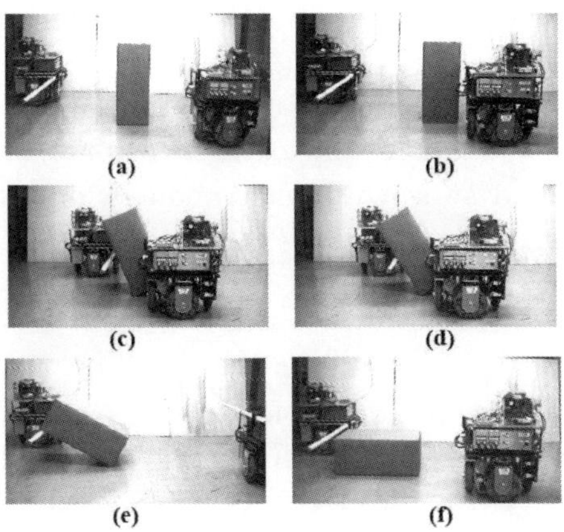

Fig. 8.4 Experimental Results of tumbling a rectangle with two robots [83].

- **Robots soccer** is considered as a dynamical and adversarial application. A ball moves and robots should act to catch and shoot it. Robots of one team should cooperate to compete with the opponent team. This application requires a lot of knowledge in different research areas such as robotics, intelligent control, communication, computer and sensor technology, image processing, mechatronics, and artificial intelligence. It is an important application for robot cooperation and competition. Figure 8.5 shows the RoboCup Soccer platform league in Robocup 2010. J.-H. Kim et al. [47] explored a soccer robot system and presented two control schemes: vision-based and robot-based. They discussed both control structure and action selection mechanisms. D.-H. Lee [49] proposed a task and role selection strategy where each robot in a team selects its task and role.

Fig. 8.5 RoboCup soccer league in 2010 (www.robocup.org)(Copyright ©1998–2013).

A plethora of applications was developed for multi-robot cooperation. Nowadays, researchers explore applications in different domains such as intelligent vehicles and swarm robotics, about which we will talk in the next section.

8.3.4 *Challenges*

Distributed information engenders a main challenge in robot cooperation. Robots exchange information through communication, thus increasing cooperation vulnerability towards errors. Communication between robots requires bandwidth and solution for data dissemination. Researchers should choose a robust network architecture capable of handling connection failures and ensuring message passing. Furthermore, by sharing information, the system is confronted to issues due to the different representation of the environment that each robot can have. This is due to several reasons such as errors in localization of each robot or different sensor equipment available on different robots. Data incest remains a challenging issue for multi-robot system. Scientists should tackle this problem by developing appropriate fusion algorithms. Several research challenges still remain such as robot computational capabilities, control architecture design, organization of high number of robots...

8.4 Swarm Robotics

It is difficult to define properly a swarm robotics, due to wide range of applications. Maybe, the most appropriate definition is: "Swarm robotics is the study of how large the number of relatively simple physically embodied agents can be designed such that a desired collective behavior emerges from the local interaction among agents and between the agents and the environment" [71]. In this definition, the main characteristics of a swarm robotics are summarized: simplicity of robots, fully distributed system, scalability, robustness. Swarm Robotics are required to be characterized with specific key advantages such as:

- Parallelism: typically a big, complex task is divided in many sub-task and each unit accomplishes a given task quicker than a single robot;
- Robustness: the system is required with a high degree of fault tolerance. In practice, if some robot fails the execution of its task, the system will evolve in a novel and dynamic configuration that will reestablish the correct functioning of the system;
- Scalability: the increment of the number of devices does not degrade the performance of the whole system;
- Heterogeneousness: each unit can be characterized with specific properties that will be effectively exploited to accomplish suitable tasks;
- Flexibility: a system has to be reconfigurable in order to accomplish different tasks and execute different applications;
- Complex Tasks: generally, a single unit is not able to accomplish a complex task, whereas a swarm is able to, because of the joint capabilities of the single devices;
- Cheap Alternative: devices are simple, easy to build and cheaper than a single powerful robot.

Typically, Swarm Robotics operate based on some sense of biological inspiration [74]. In this sense, the application of Swarm Intelligence to collective robotics can be identified as "Swarm Robotics". The sense of the interaction between bio-inspiration, Swarm Intelligence and Self-organized and Distributed System can be explained through the Fig. 8.6.

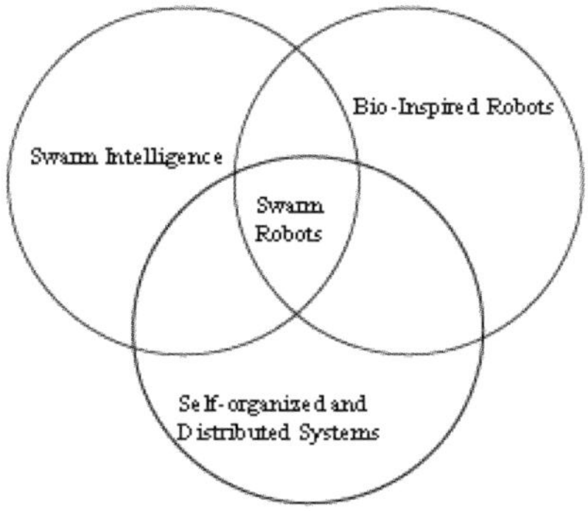

Fig. 8.6 Swarm Robotics as the intersection between bio-inspired systems, robots Swarm Intelligence and Self-Organized and Distributed Systems [63].

From an historical point of view, the first experiments on systems of robots that could be identified as Swarm Robotics, were realized in late 1940s. Grey Walter and his team showed a system of simple robots interacting in a seemingly social manner and by exhibiting "complex behavior" [24], but Swarm Robotics becomes an active field of research only in the 1990's. G. Beni [9] introduced the concept of Swarm Robotics by discussing cellular robotics systems. In the 1990's Deneubourg et al. introduced the concept of stigmergy in robots that behave like ants [21], [7]. Since then, numerous researchers have developed collective and self-organized systems [18] and have introduced robots' behaviors inspired by insects' social organization [53], [25], [16].

8.4.1 *Classification of swarm robotics*

Different types of classification have been proposed for Swarm Robotics. In [1] authors propose a taxonomy and classify existing studies. Specifically, they split existing studies into the most important research directions. The five fields they identify are: modeling, behavior design, communication, analytical studies and problems. The taxonomy is summarized in Fig. 8.7.

Concerning *modeling*, authors found that modeling is a very suitable method for Swarm Robotics. In fact, there are some risks related to the

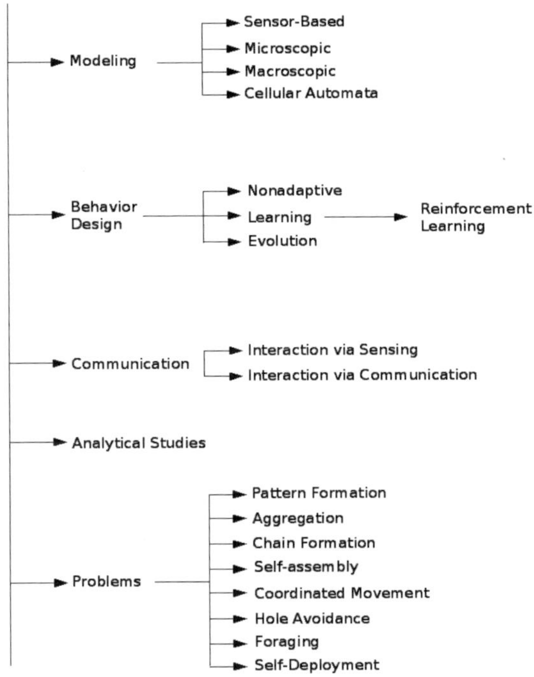

Fig. 8.7 Classification of Swarm Robotics literature [1].

robots that require a human to follow the experiments. Typically, to validate results, a high number of experiments is required and simulation and modeling of the experiments seem to be an effective way to make the system work. Another important aspect related to modeling in Swarm Robotics is scalability. Generally, demonstration of scalability of some control algorithm requires hundreds of robots. Costs related to the use of such a number of robots could be prohibitive and modeling could become the only viable solution.

In a biological system, individuals may fine-tune their behaviors in their lifetime. In practice, they learn how to survive and to stay better when external conditions change. In Swarm Robotics, researchers considered the *behavioral adaptation* to control large number of robots to accomplish a task collectively.

Communication is sub-divided into three types. The first kind is via sensing and represents the simplest type of communication based on the capacity of a robot to distinguish between other robots and the objects in the environment. When robots use interaction via the environment, they consider it as a communication medium (i.e., pheromones used from ants). Interaction via communication involves explicit communication through direct messages.

Analytical studies include studies that contribute to the theoretical understanding of swarm systems. In this category, methods for solution of different problems can be included. Furthermore, mathematical tools that allow a deeper comprehension of the details of Swarm Robotics systems can be considered as part of analytical studies.

The last point of the taxonomy formulated in [1] is the *problem* axis, where authors individuate general problems that have been investigated in Swarm Robotics. A researcher could find useful the individuation of this point when he tries to solve a specific problem and can try to make its problem matchable with a more general problem already faced in literature.

In [36] authors classified Swarm Robotics literature in different domains. They suggest a classification based on the characteristics of the swarm as a whole rather than the architectural characteristics of individual robots. Authors individuate domains like communication range, communication topology, swarm size, communication bandwidth, etc.

In [82] authors present a survey of cooperative robotics in a hierarchical way, as we have already mentioned in the previous section. They individuate five main axes: group architecture, resource conflicts, origins of cooperation, learning and geometric problems. Group architecture represents the necessary infrastructure where the cooperative behavior must rely. Resource conflicts is strictly related to the communication of the robots, the management of the shared environment, etc. Origin of cooperation means how cooperative behavior is achieved and actuated. Learning axis is strictly related to adaptability and flexibility that represent essential traits in a task-solving Swarm Robotics. These first four axes are mostly related to the cooperative aspects of Swarm Robotics. The fifth axis individuated by the authors is defined as geometric problems and covers research issues tied to the embedding of robot tasks in a two- or three-dimensional world.

In [44] authors survey existing works on modeling collective behavior of robot swarms with macroscopic models. Specifically, they consider very simple robots that can be represented as stochastic Markov processes. A macroscopic model describes the collective behavior of the robotic swarm. The choice of macroscopic vs microscopic models, by taking into account the behavior of some *average* quantity that represents the system, is related to the inherent simplicity and the analytical tractability of such macroscopic descriptions.

A more recent and interesting classification of Swarm Robotics is given in [13]. Authors propose a classification based on two taxonomies as shown in Fig. 8.8.

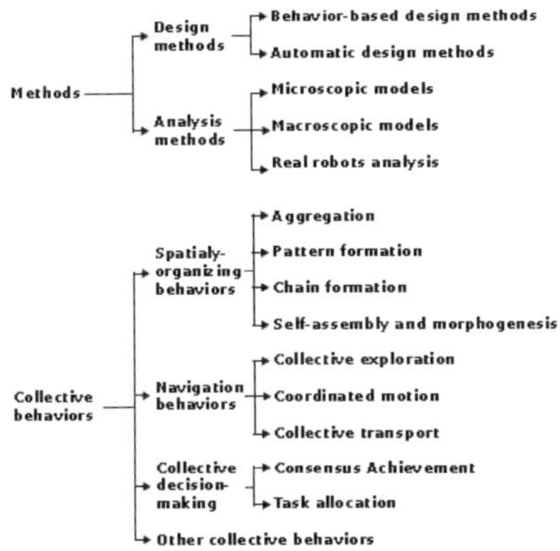

Fig. 8.8 Taxonomies for Swarm Robotics [13].

Authors classify literature regarding Swarm Robotics according to the methods used to design or analyze the swarms and its main collective behaviors.

8.4.2 *Applications of swarm robotics*

Concerning potential domains where Swarm Robotics can be applied, various scenarios can be individuated:

- *Foraging* - In this scenario, a robot is able to collect the objects and deliver them to some predefined location. This scenario requires many fundamental skills from a Swarm Robotic system, such as collective exploration, efficient task allocation, etc. It seems that the first contribution in terms of implementation of foraging using a group of real robots is given in [62]. In [57], Mataric considers a formulation of reinforcement learning in a concurrent multi-robot learning domain. In order to validate the fach, the author proposes an experiment involving four mobile robots learning a foraging task. Sugawara et al. investigate the collective and cooperative behavior of interacting agents [42].

The task assigned to the swarm consists into pick up and collect pucks distributed in a field. This task is reformulated as a problem of ants foraging even if the movement and interaction of agents (robots) are more simplified than ants. In [31], Ducatelle et al. study self-organization of heterogeneous swarms robotic to solve a specific task. Specifically, they consider two swarms that need to mutually adapt to each other and the swarm, as a whole, has to solve the task. Their work is related to the problem on self-organized foraging, where robots have to optimize a path to follow back and forth between a source and a target [34].

- *Aggregation* - Aggregation is one of the fundamental behaviors of swarm in nature. In Swarm Robotics, self-organized aggregation is required to form a robot cluster and is a very common goal but the approaches are very diverse. In [27], authors consider an evolving neural network with 12 neurons to reach robot aggregation. Additionally, aggregation is a requisite for Swarm Robotics behaviors such as self-assembly and pattern formation [78]. In [12], authors show how the spatial separation of two conflictive spots affects the cooperation behavior. In [5], authors investigate aggregation behavior as a case, and systematically studies the performance and the scalability of aggregation behaviors of perceptron controllers evolved for a simulated Swarm Robotic system with different parameter settings. Baldassarre et al. successfully evolved controllers for a swarm of robots to aggregate and move towards a light source in a clustered formation [35].
- *Clustering and Sorting* - Clustering and sorting are mostly influenced by the nest building behavior of termites and wasps. Distributed clustering, and more recently sorting, by a swarm of robots have served as benchmarks for swarm intelligence based robotics [29]. In [2] a new method for distributed object sorting by a swarm of robots is introduced. In this work it is shown how an unloaded agent seeks an isolated object to pick up, and an agent already carrying an object seeks an existing cluster of the same type to deposit its load. Authors employ only on-board sensing. In [23] authors use the concept of spatial awareness to accomplish the cluster task and support task allocation that are spatially differentiated.
- *Exploration* - Exploration of an unknown environment is a fundamental issue in mobile robotics. One of the main advantages in the usage of multiple robots instead of a single one lies on the speed of convergence of the exploration process, accuracy of the solution, and fault tolerance. Most significant research topics in multi-robot-exploration

are task sharing and navigation. A useful contribution in terms of multi-robot exploration is given in [54]. The exploration method proposed by these authors minimizes the overall exploration time, making it possible to efficiently localize fire sources. In GUARDIANS (Group of Unmanned Assistant Robots Deployed In Aggregative Navigation by Scent)FP6, EU funded project, there is a group of robots whose task is the exploration of the unknown operative environment. In [19] authors propose two techniques based on Particle Swarm Optimization and Darwinian Particle Optimization to perform in an effective way exploration task, by explicitly taking obstacle avoidance into account.

8.5 Conclusion

In the first part of this chapter, we tried to express the essence of what Swarm Intelligence is, namely: "A single ant or bee is not smart, but their colonies are. The study of Swarm Intelligence is providing insights that can help humans manage complex systems, from truck routing to military robots" [60]. After considering the main applications of Swarm Intelligence itself, we presented the most important results about cooperation of robots, trying to analyze in a critical fashion the issues and challenges related to this field and presenting the main advantages. The last contribution of this chapter is represented by the synergic "fusion" of Swarm Intelligence and cooperation among robots, that we referred as Swarm Robotics. Specifically, we presented some interesting taxonomies and typical applications of this new field of research.

References

[1] A., Baynidir and E., Sahin (2007). A review of studies in swarm robotics, *Turkish journal of electrical engineering and computer science* **15**, pp. 115–147.

[2] A., Vardy (2012). Accelerated patch sorting by a robotic swarm. in *Ninth Conference on Computer and Robot Vision (CRV)*, pp. 314–321.

[3] AlRashidi, M. and El-Hawary, M. (2009). Applications of computational intelligence techniques for solving the revived optimal power flow problem, *Electric Power Systems Research* **79**, 4, pp. 694–702, doi:10.1016/j.epsr.2008.10.004.

[4] Andersson, L. and Nygards, J. (2003). On sensor fusion between a pair of heterogeneous robots, in *Proceedings of the 6th International Conference on Information Fusion, FUSION 2003* (Cairns, Australia), pp. 1287–1294.

[5] Baheci, I. and Sahin, E. (2005), Evolving aggregation behaviours for swarm robotic systems: a systematic case study. In *Proceedings of IEEE swarm intelligence symposium.*, pp. 333–340.

[6] Bautin, A., Simonin, O. and Charpillet, F. (2011). Towards a communication free coordination for multi-robot exploration, in *6th National Conference on Control Architectures of Robots, CAR 2011* (Grenoble, France).

[7] Beckers, R., Holland, O. and Deneubourg, J.-L. (1994). From local actions to global tasks: Stigmergy and collective robotics, in *Artificial Life*.

[8] Belbachir, A., Lacroix, S., Ingrand, F., Perrier, M. and Opderbecke, J. (2010). Cooperative-adaptive algorithms for targets localization in underwater environment, in *IEEE-OES Autonomous Underwater Vehicles* (Monterey, CA).

[9] Beni, G. and Wang, J. (1989). Swarm intelligence in cellular robotic systems, in *Proceedings of NATO Advanced Workshop on Robots and Biological Systems* (Italy).

[10] Bishop, J. (1989). Stochastic searching networks, in *1st IEE Conf. Artificial Neural Networks* (Reading University, UK).

[11] Blum, C. and Merkle, D. (2008). *Swarm Intelligence: Introduction and Applications* (Natural Computing Series. Springer).

[12] Bodi, M., Thenius, R., Schmickl, T. and Crailsheim, K. (2011). How two cooperating robot swarms are affected by two conflictive aggregation spots, in *Proceedings of the 10th European conference on Advances in artificial life: Darwin meets von Neumann - Volume Part II*, ECAL'09 (Springer-Verlag, Berlin, Heidelberg), ISBN 978-3-642-21313-7, pp. 367–374.

[13] Brambilla, M., Pinciroli, C., Birattari, M. and Dorigo, M. (2012). Property-driven design for swarm robotics, in *IAAMAS*, pp. 139–146.

[14] Cai, Y. and Yang, S. X. (2012). A survey on multi-robot systems, in *World Automation Congress (WAC), 2012* (Puerto Vallarta, Mexico).

[15] Camazine, S., Deneubourg, J.-L., Franks, N., Sneyd, J., Theraulaz, G. and Bonabeau, E. (2001). *Self-organization in biological systems* (Princeton University Press).

[16] Campo, A., Gutiérrez, A., Nouyan, S., Pinciroli, C., Longchamp, V., Garnier, S. and Dorigo, M. (2010). Artificial pheromone for path selection by a foraging swarm of robots, *Biological Cybernetics* **103**, 5, pp. 339–352, doi: http://dx.doi.org/10.1007/s00422-010-0402-x.

[17] Chen, L., Arambel, P. O. and Mehra, R. K. (2002). Estimation under unknown correlation: covariance intersection revisited, *IEEE Transactions on Automatic Control* **47**, pp. 1879–1882.

[18] Correll, N. and Martinoli, A. (2006). System identification of self-organizing robotics swarms, in *DARS*, pp. 31–40.

[19] Couceiro, M. (2011). A novel multi-robot exploration approach based on particle swarm optimization algorithms. in *In Proceedings of IEEE International Symposium on Safety, Security, and Rescue Robotics (SSRR)*, pp. 327–332.

[20] Cui, Y. A. and Dai, Y. J. (2011). Robots active linking in unknown environment, *Advanced Materials Research* **186**, pp. 84–88.

[21] Deneubourg, J. L., Goss, S., Franks, N., Sendova-Franks, A., Detrain, C. and Chrétien, L. (1991). The dynamics of collective sorting: Robots-like ants and ants-like robots. from animal to animats, in *Artificial Life*, pp. 356–363.

[22] Denoeux, T. (2008). Conjunctive and disjunctive combination of belief functions induced by nondistinct bodies of evidence, *Artificial Intelligence* **172**, pp. 234–264, doi:10.1016/j.artint.2007.05.008.

[23] Di Caro, G. A., Ducatelle, F. and Gambardella, L. (2012). A fully distributed communication-based approach for spatial clustering in robotic swarms, in *Proceedings of the 2nd Autonomous Robots and Multirobot Systems Workshop (ARMS), affiliated with the 11th International Conference on Autonomous Agents and Multiagent Systems (AAMAS)* (Valencia, Spain, June 5), pp. 153–171.

[24] Dorf, R. (1990). *Concise international encyclopaedia of robotics: Applications and automation* (Wiley-Interscience).

[25] Dorigo, M. (ed.) (1996). Special Issue on Learning Autonomous Robots, *IEEE Transactions on Systems, Man, and Cybernetics – Part B*, 26:361–506 (IEEE Press).

[26] Dorigo, M. and Stützle, T. (2004). *Ant Colony Optimization* (MIT Press, Cambridge, USA).

[27] Dorigo, M., Trianni, V., Şahin, E., Groß, R., Labella, T. H., Baldassarre, G., Nolfi, S., Deneubourg, J.-L., Mondada, F., Floreano, D. and Gambardella, L. M. (2004). Evolving self-organizing behaviors for a swarm-bot, *Autonomous Robots* **17(2-3)**, pp. 223–245.

[28] Ducourthial, B., Cherfaoui, V. and Denoeux, T. (2012). Self-stabilizing distributed data fusion, in *14th International Symposium on Stabilization, Safety, and Security of Distributed Systems (SSS 2012)*, Vol. 7596 (Toronto), pp. 148–162.

[29] Dorigo, M., Bonabeau, E. and Theraulaz, G. (1999). *Swarm Intelligence: From Natural to Artificial Systems* (Oxford University Press, New York, USA).

[30] Eslami, M., Shareef, H., Khajehzadeh, M. and Mohamed, A. (2012). An effective particle swarm optimization for global optimization, *Computational Intelligence and Intelligent Systems*, pp. 267–274doi:10.1007/978-3-642-34289-9_30.

[31] Ducatelle, F., Di Caro, G. A., Förster, A. and Gambardella, L. (2010). Mobile stigmergic markers for navigation in a heterogeneous robotic swarm, *Swarm Intelligence, Proceedings of the 7th International ANTS Conference.*

[32] Fox, D., Burgard, W., Kruppa, H. and Thrun, S. (2000). A probabilistic approach to collaborative multi-robot localization, *Autonomous Robots* **8**, pp. 325–344.

[33] Franchi, A., Oriolo, G. and Stegagno, P. (2009). Mutual localization in a multi-robot system with anonymous relative position measures, in *2009 IEEE/RSJ Int. Conf. on Intelligent Robots and Systems* (St. Louis, MO), pp. 3974–3980.

[34] Fujisawa, R., Dobata, S., Kubota, D., Imamura, H. and Matsuno, F. (2008). Dependency by concentration of pheromone trail for multiple robots, *Proc.*

of the 6th Int. Conf. on Ant Colony Optimization and Swarm Intelligence (ANTS).

[35] Baldassare, G., Nolfi, S. and Parisi, D. (2003). Evolving mobile robots able to display collective behavior, in *Artificial Life*, Vol. 9(3), pp. 255–267.

[36] Dudek, G., Jerkin, M. and Wilkes, D. A. (1993). Taxonomy of swarm robots, *IEEE Int. Conf. on Intelligent robots and systems*, pp. 441–447.

[37] H. Zhu, K. E., C. Pu and Gu, J. (2009). Euclidean particle swarm optimization, in *In Intelligent Networks and Intelligent Systems (ICINIS).*

[38] Hidaka, Y. S., Mourikis, A. I. and Roumeliotis, S. I. (2005). Optimal formations for cooperative localization of mobile robots, in *Proc. of the IEEE International Conference on Robotics and Automation* (Barcelona, Spain), pp. 4137–4142.

[39] Eberhart, R. C., Kennedy, J. and Shi, Y. (2004). *Swarm Intelligence* (Morgan Kaufmann Publishers).

[40] Jia, Y. and Xi, N. (2011). Coordinated formation control for multi-robot systems with communication constraints, in *2011 IEEE/ASME International Conference on Advanced Intelligent Mechatronics (AIM2011)* (Budapest, Hungary).

[41] Julier, S. and Uhlmann, J. (2001). *Handbook of Multidimensional Data Fusion, Chap. 12*, chap. General decentralized data fusion with covariance intersection (CI) (Hall D.and Llians J. eds, CRC Press).

[42] Sugawara, K. and Sano, M. (1997). Cooperative acceleration of task performance: Foraging behavior of interacting multi-robots system. *Physica*, pp. 343–354.

[43] Nasuto, S. J., De Meyer, K. and Bishop, J. M. (2006). *Stochastic Diffusion Search: partial function evaluation in swarm intelligence dynamic optimisation*, Vol. 31 (Abraham A. and Grosan C. and Ramos, V., eds, Springer Verlag, Germany).

[44] Galstyan, A., Lerman, K. and Martinoli, A. (2005). *A review of probabilistic macroscopic models for swarm robotic systems*, number 3342 in LNCS (Springer-Verlag, Berlin Heidelberg).

[45] Karam, N., Chausse, F., Aufrre, R. and Chapuis, R. (2006). Localization of a group of communicating vehicles by state exchange, in *Proceedings of the IEEE/RSJ International Conference on Intelligent Robots and Systems* (Beijing, China), pp. 2816–2821.

[46] Kennedy, J. and Eberhart, R. (1995). Particle swarm optimization, in *In Proceedings of the Fourth IEEE International Conference on Neural Networks* (Perth, Australia), pp. 1942–1948.

[47] Kim, J.-H., Shim, H.-S., Jung, M.-J., Kim, H.-S. and Prahlad., V. (1997). Cooperative multi-agent robotic systems: From the robot-soccer perspective, in *Proc. of MiroSot97 Workshop*, pp. 3–14.

[48] Lacroix, S. and Besnerais, G. L. (2011). Issues in cooperative air/ground robotic systems, *Robotics Research*, pp. 421–432.

[49] Lee, D.-H., Na, K.-I. and Kim, J.-H. (2010). Task and role selection strategy for multi-robot cooperation in robot soccer, *Trends in Intelligent Robotics Communications in Computer and Information Science* **103**, pp. 170–177.

[50] Lee, J., Jo, K. and Jang, C. (2007). Cooperative localization for multirobot incorporative proprioceptive/exteroceptive position sensors, in *6th Int. Conf Field and Service Robotics* (Chamonix, France).

[51] Leitner, J. (2009). A survey of multi-robot cooperation in space, *Advanced Technologies for Enhanced Quality of Life*, pp. 144–151.

[52] Lopez, J., Perez, D. and Zalama, E. (2011). A framework for building mobile single and multi-robot applications, *Robotics and Autonomous Systems* **59**, 3-4, pp. 151–162.

[53] Kriger, M., Billeter, J.-B. and Keller, L. (2000). Ant like task allocation and recruitment in cooperative behaviour, in *Nature*, ISBN 406(6799).

[54] Marjovi, A., Nunes, J., Marques, L. and Almeida, A. (2009). Multi-robot exploration and fire searching, in *IEEE/RSJ International Conference on Intelligent Robots and Systems (IROS)*.

[55] Martinelli, A., Pont, F. and Siegwart, R. (2005). Multi-robot localization using relative observations, in *Proceedings of the 2005 IEEE International Conference on Robotics and Automation* (Barcelona, Spain), pp. 2808–2813.

[56] Mas, I. and Kitts, C. (2012). Object manipulation using cooperative mobile multi-robot systems, in *Proceedings of the World Congress on Engineering and Computer Science, WCECS 2012* (San Francisco, USA).

[57] Mataric, M. (1997). Reinforcement learning in the multi-robot domain. *Autonomos robots*, pp. 73–83.

[58] Meyer, K. D. (2004). Foundations of stochastic diffusion search, *Culture* .

[59] Miettinen, K. (ed.) (1999). *Evolutionary Algorithms in Engineering and Computer Science: Recent Advances in Genetic Algorithms, Evolution Strategies, Evolutionary Programming, GE* (John Wiley & Sons, Inc., New York, NY, USA), ISBN 0471999024.

[60] Miller, P. (2007). The genius of swarm, *National Geographic Magazine* .

[61] Mitchell, H. (2007). *Multisensor Data Fusion: An introduction* (Springer).

[62] Marjonovic, M. J and Mataric, M. J. (1993). Synthesizing complex behaviours by composing simple primitives. in *European conference on Artificial life*, Vol. 2, pp. 698–707.

[63] Pujol, A. O. and Muniganti, P. (2010). A survey on mathematical models of swarm robotics. in *Physical Agents Network, XI Workshop of Physical Agents*.

[64] Parker, L. E. (2003). Current research in multirobot systems, *Journal of Artificial Life and Robotics* **7**, pp. 1–5.

[65] Pimentel, B. S. and Campos, M. F. M. (2003). Cooperative communication in ad hoc networked mobile robots, in *Proceedings of the 2003 IEEE/RSJ Intl. Conference on Intelligent Robots and Systems* (Las Vegas, Nevada), pp. 2876–2881.

[66] Prorok, A., Bahr, A. and Martinoli, A. (2012). Low-cost collaborative localization for large-scale multi-robot systems, in *IEEE International Conference on Robotics and Automation (ICRA)* (RiverCentre, Saint Paul, Minnesota, USA).

[67] Kennedy, J., Poli R. and Blackwell, T. (2007). Particle swarm optimization, *Swarm Intelligence* **1**, 1, pp. 33–57.

[68] Rigatos, G. (2008). Distributed motion planning and sensor fusion for cooperative behavior of mobile robots, in *Proceedings of the EURON/IARP International Workshop on Robotics for Risky Interventions and Surveillance of the Environment* (Benicassim, Spain).

[69] Roumeliotis, S. and Bekey, G. A. (2002). Distributed multirobot localization, *IEEE Transactions on Robotics and Automation* **18**, 5, pp. 781–795.

[70] Gautrai, J., Garnier, S. and Theraulaz, G. (2007). The biological principles of swarm intelligence, *Swarm Intelligence* **1**.

[71] Sahin, E. (2005). Swarm robotics: From source of inspiration to domains of applications in swarm robotics, in *Springer Verlag*, pp. 10–20.

[72] Santos, J. and Lima, P. (2010). Multi-robot cooperative object localization-decentralized bayesian approach, in *Robot Soccer World Cup XIII, Lecture Notes in Artificial Intelligence*, Vol. 5949 (Graz, Austria), pp. 332–343.

[73] Seeley, T. (2002). When is self-organization used in biological systems? *Biological Bulletin*, pp. 314–318.

[74] Sharkey, A. (2006). The application of swarm intelligence to collective robots, in *Advances in Applied Artificial Intelligence* (John Fulcher, Idea Group Publishing), pp. 157–185.

[75] Siciliano, B. and Khatib, O. (eds.) (2008). *Springer Handbook of Robotics* (Springer).

[76] Silva, J., Lau, N., Rodrigues, J., Azevedo, J. L. and Neves, A. J. R. (2009). Sensor and information fusion applied to a robotic soccer team, in Springer (ed.), *Robocup 2009: robot soccer worls cup XIII. Lecture notes in artificial intelligence.*

[77] Song, K.-T., Tsai, C.-Y. and Huang, C.-H. C. (2008). Multi-robot cooperative sensing and localization, in *IEEE International Conference on Automation and Logistics, ICAL 2008.*, pp. 431–436.

[78] Soysal O., S. E., Bahceci E. (2007). Aggregation in swarm robotic systems: Evolution and probabilistic control. *Turkish Journal Electrical Engineering* **15(2)**, pp. 199–225.

[79] Verret, S. (2005). Current state of the art in multirobot systems, Technical report, Defence R&D Canada, Suffield.

[80] Wang, S. and Tan, M. (2000). Multi-robot cooperation and data fusion in map-building, in *Proceedings of the* 3rd *World Congress on Intelligent Control and Automation* (Hefei, P.R. China), pp. 1261–1265.

[81] Wang, T., Dang, Q. and Pan, P. (2011). A predict-fuzzy logic communication approach for multi robotic cooperation and competiton, *Journal Of Communications* **6**, 3, pp. 225–231.

[82] Cao, Y., Fukunaga, A. and Kahng, A. (1997). Cooperative mobile robotics: Antecedents and directions, *Autonomous robots* **4(1)**, pp. 7–27.

[83] Yamashita, A., Fukuchi, M., Ota, J., Arai, T. and Asma, H. (2000). Motion planning for cooperative transportation of a large object by multiple mobile robots in a 3d environment, in *Proccedings of the 2000 IEEE International Conference on Robotics and Automation* (San Francisco, CA), pp. 3144–3151.

[84] Yan, Y. and Zhenmin, T. (2009). Control architecture for autonomous multi-robot system: Survey and analysis, in *Second International Conference on Intelligent Computation Technology and Automation, 2009. ICICTA'09*, Vol. 4 (USA), pp. 376–379.

[85] Zhao, X., Luo, Q. and Han, B. (2008). Survey on robot multi-sensor information fusion technology, in *Proceedings of the 7th World Congress on Intelligent Control and Automayion* (Chongqing, China), pp. 5014–5018.

[86] Zoghby, N. E., Cherfaoui, V., Ducourthial, B. and Denoeux, T. (2012). Distributed data fusion for detecting sybil attack in vanets, in *Belief Functions: Theory and Applications*, Vol. 164, pp. 351–358.

Chapter 9

Localization in Wireless Sensor Networks

Roudy Dagher[1,2] and Roberto Quilez[2]

[1] *Etineo, France,* [2] *Inria Lille – Nord Europe, France*

Abstract *With the proliferation of Wireless Sensor Networks (WSN) applications, knowing the node current location have become a crucial requirement. Location awareness enables various applications from object tracking to event monitoring, and also supports core network services such as: routing, topology control, coverage, boundary detection and clustering. Therefore, WSN localization have become an important area that attracted significant research interest. In the most common case, position related parameters are first extracted from the received measurements, and then used in a second step for estimating the position of the tracked node by means of a specific algorithm. From this perspective, this chapter is intended to provide an overview of the major localization techniques, in order to provide the reader with the necessary inputs to quickly understand the state-of-the-art and/or apply these techniques to localization problems such as robot networks. We first review the most common measurement techniques, and study their theoretical accuracy limits in terms of Cramer-Rao lower bounds. Secondly, we classify the main localization algorithms, taking those measurements as input in order to provide an estimated position of the tracked node(s).*

9.1 Introduction

Recent technological advances in micro-electronics, digital electronics and wireless communication, have made possible the development of low-cost, low-power, multi-functional and highly integrated sensor nodes that are able to communicate in a wireless ad-hoc fashion over short distances [3]. These tiny nodes, typically equipped with processing, sensing, power management and communication capabilities collaborate to form a Wireless Sensor Network (WSN). Sensed data is typically sent over the network, in a multi-hop manner, to a control center either directly or via a *base station/sink*. The main constraints in such networks are the limited amount of energy and computing resources of the nodes.

With the significant development and deployment of WSN, associating the sensed data with its physical location becomes a crucial requirement. Knowing the node's location enables a myriad of location-based applications such as object tracking, environment monitoring, intrusion detection, and habitat monitoring [73] [25]. Location estimation also supports core network services such as: routing, topology control, coverage, boundary detection and clustering [43].

Localization is defined as the process of obtaining a node location with respect to a set of known reference positions. It is also referred to as *location estimation* or *positioning*. Nodes at reference positions are called *anchor nodes*[1], and nodes with unknown positions are called *tracked nodes*[2]. Based on reference positions of a few anchor nodes in the network, and inter-node measurements such as range and connectivity, localization algorithms estimate the position of a tracked node in the network. Depending on targeted applications, the coordinate system may be global or local (e.g. habitat monitoring).

In the most common case, position related parameters are first extracted from the received measurements, and then used in a second step for estimating the position of the tracked node by means of a specific approach: fingerprinting, geometric or statistical methods [20]. The used technique highly depends on the application's requirements and challenges:

- *Environment*
  The environment where a WSN is deployed may be challenging, as localization performance is affected by multipath and non-line-of-sight

[1]also referred to as reference node, beacon device, base station, etc.

[2]also referred to as non-anchor node, target node, blindfolded device, mobile station etc.

(NLOS) propagation. Environment variability is typically due to: presence of obstacles, metallic environments acting as wave guides, interference, etc.

- *Complexity*
  In the context of WSN, nodes are typically battery-powered with limited computing power and memory. Therefore, it may not be feasible to implement complex localization algorithms. However, in some cases the base station may have advanced computing capabilities and act as a localization server for the application.
- *Accuracy*
  Coarse-grained accuracy of several meters may be sufficient for patient tracking inside a hospital, and may be addressed by simple low-cost Zigbee-based solutions [17]. Conversely, fine-grained accuracy usually requires specialized hardware such as Ultra-wideband (UWB) [22].
- *Scalability*
  Scalability of a localization algorithm determines how well it accommodates as the number of nodes increases and the coverage area is expanded. This metric is very important in dense networks.
- *Latency*
  Depending on the tracked object dynamics, the latency to determine its location might be a big concern. It should be considered with respect to other layers such as the Medium Access Control (MAC) layer for channel access latency.
- *Dependability*
  The system should be able to keep operating even if some anchor nodes are faulty. This is referred to as system *fault tolerance*. In [51] a WSN localization system with error detection/correction is presented. Another issue to consider is *network lifetime*, mainly in the case of battery-powered nodes.

In brief, the choice of a sensor network localization technique often involves a trade-off among the above-listed constraints in order to suit the requirements of the targeted application(s). In essence, these challenges make localization in wireless sensor networks unique and intriguing. This chapter is intended to provide an overview of the major techniques that have been widely used for WSNs localization system. Based on the referenced material, a special effort has been made to broadly classify the different localization aspects in order to provide a starting block for this topic. The remainder of this chapter is organized as depicted in Fig. 9.1. First section

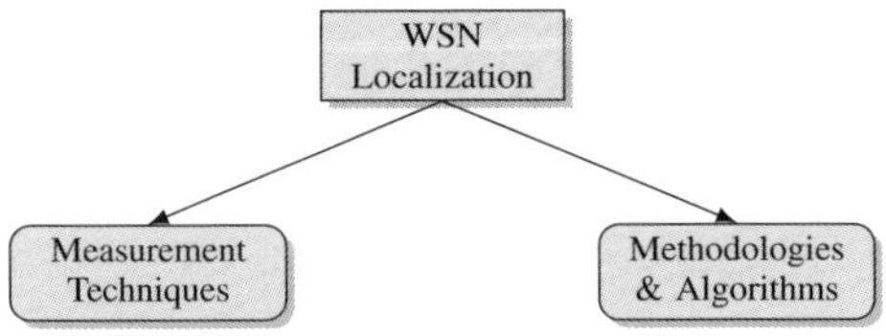

Fig. 9.1 Chapter overview.

presents the measurement techniques and their theoretical accuracy limits in terms of Cramer-Rao lower bounds. The second section covers the localization theory, strategies and algorithms taking those measurements as input in order to provide an estimated position of the tracked node(s).

9.2 Measurement Techniques

Position related parameters estimation is the first step of WSN localization. This estimation often relies on physical measurements, depending on the available hardware capabilities. On the other hand, network related measurements such as hop count, or neighborhood information can lead to coarse-grained localization that may be sufficient in dense networks. Figure 9.2 gives an overview of these measurement techniques. It is the type of measurements employed and the corresponding precision that fundamentally determine the estimation accuracy of a localization system and the localization algorithm being implemented by this system.

In the case of physical measurements, a result of estimation theory can be used to bound the localization error: the *Cramer-Rao lower bound* (CRLB) [30, 82]. This theoretical bound gives the best performance that can be achieved by an unbiased location estimator. If $\hat{\theta}$ is an unbiased estimator of an unknown parameter θ, then its covariance matrix $Cov(\hat{\theta})$ is bounded by the CRLB as

$$Cov(\hat{\theta}) \geq \left\{ -E\left[\nabla_\theta(\nabla_\theta \ln f(X|\theta))\right] \right\}^{-1} \tag{9.1}$$

where X is the random observation vector with probability density function $f(X|\theta))$, $E[\cdot]$ indicates the expected value, and ∇_θ is the gradient operator with respect to θ. Note that the CRLB is independent of the estimation method, and only depends on the statistical model of the observations. Therefore, the CRLB can serve as a benchmark for localization algorithms.

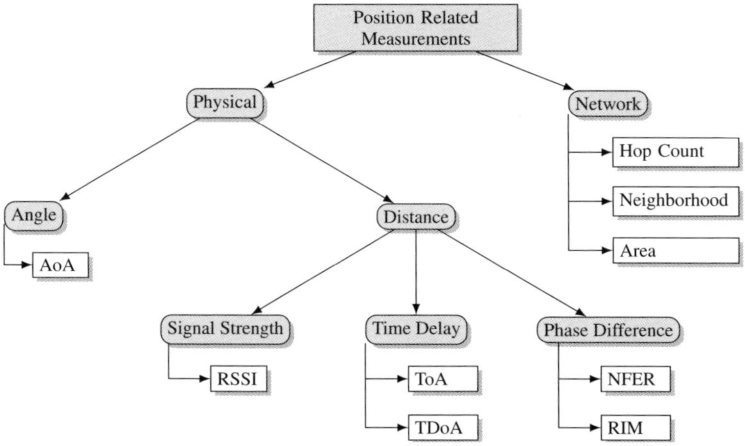

Fig. 9.2 Measurement techniques overview.

In the case where θ is scalar, the CRLB in Eq. (9.1) becomes

$$\sigma_{\hat{\theta}}^2 \geq \frac{1}{-E\left[\frac{\partial^2 \ln f(X|\theta)}{\partial \theta^2}\right]} = \frac{1}{-\int_{\mathbb{R}} \frac{\partial^2 \ln f(X|\theta)}{\partial \theta^2} f(X|\theta) \mathrm{d}X}. \tag{9.2}$$

9.2.1 *Physical measurements*

Physical measurements can be broadly classified into three categories according to the measurement type: angle measurements, distance related measurements and network related measurements.

9.2.1.1 *Angle measurements*

The angle or bearing relative to reference nodes is measured by estimating the angle of arrival (AoA) parameter between the tracked node and reference nodes. Given the angle measurements, the location of the tracked node may be determined by triangulation[3] [67]. The AoA measurement is commonly made available by the use of directional antennas or antenna arrays[4], by measuring the phase difference between the signal received by adjacent antenna elements $\Delta\Phi = 2\pi \frac{\Delta \sin \alpha}{\lambda}$ with Δ the inter-element spacing of the N_a elements antenna array, and λ the wavelength. In order to

[3]The use of triangulation to estimate distances goes back to antiquity: Thales similar triangles to estimate the height of the pyramids, distances to ships at sea as seen from a cliff, etc.

[4]Another technique uses receiver antenna's amplitude response [47].

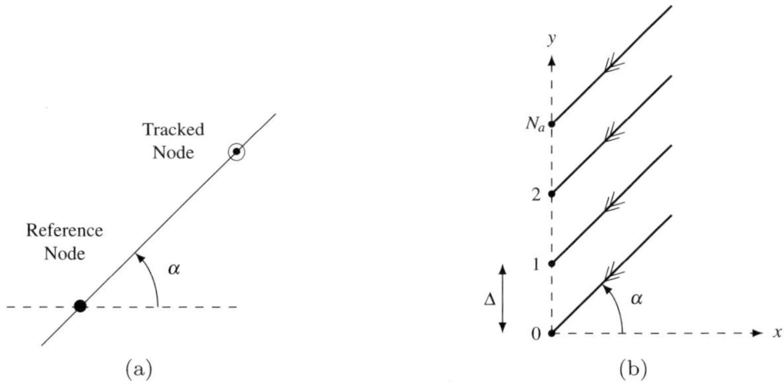

Fig. 9.3 AoA measurement with antenna array. (a) AoA definition. (b) Antenna Array.

understand the accuracy limit with this type of measurements, consider a uniform antenna as in Fig. 9.3. Under the assumption that the signal, with effective bandwidth β, arrives at the speed of light c at each antenna element via a single path, with the same signal to noise ratio (SNR) for all elements, the CRLB for the variance of an unbiased AoA estimate α can be expressed by [22]:

$$\sqrt{Var\{\hat{\alpha}\}} \geq \frac{\sqrt{3}c}{\sqrt{2}\pi\sqrt{\mathrm{SNR}}\beta\sqrt{N_a\left(N_a^2-1\right)}\Delta\cos\alpha}. \tag{9.3}$$

Equation (9.3) states that the AoA accuracy increases with the SNR, effective bandwidth, and the array size. Finally, the best accuracy is obtained when the signal direction and the antenna line are perpendicular i.e., $\alpha = 0$.

For a detailed overview on AoA measurements, refer to [47].

9.2.1.2 *Distance measurements*

By using the distance of the tracked node to several reference nodes, the position of the tracked node can be computed using the multilateration method [73].

In order to estimate the distance, several ranging techniques have been developed [47]. Among them, the Received Signal Strength (RSS) based, and the time based ranging are the most popular. A less popular technique, is the Near Field Electromagnetic Ranging (NFER) that exploits specific near-field properties of radio waves for ranging purposes. Yet another less adopted technique, the Radio Interferometric Positioning (RIPS) that applies *interferometry* to radio waves in WSN.

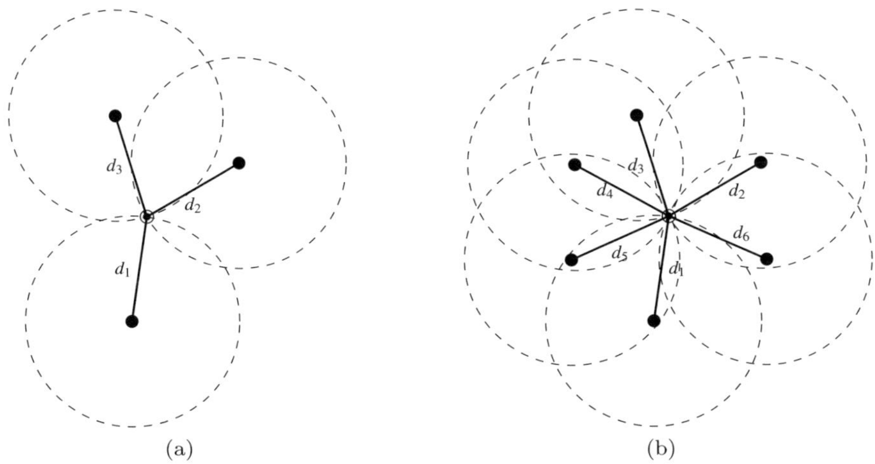

Fig. 9.4 Distance measurements and multilateration (Ranging circles). (a) Trilateration. (b) Multilateration.

Received Signal Strength This technique is based on the received signal strength indicator (RSSI), a standard feature found in most of the current off-the-shelf devices. The most typical RSS systems are based on propagation loss equations [66], given by the following log-normal shadowing model[5]

$$P_r(d)[dBm] = P_0(d_0)[dBm] - 10n_p log_{10}\frac{d}{d_0} + X_\sigma \tag{9.4}$$

where:

- $P_r(d)[dBm]$: the received power in dB milliwatts at distance d from the transmitter.
- $P_0(d_0)[dBm]$: the reference power in dB milliwatts at a reference distance d_0 from the transmitter.
- n_p : the path loss exponent that measures the rate at which the received signal strength decays with distance. Example of values: 2.0 in free space, 1.6 to 1.8 inside a building [66].
- X_σ : a zero-mean normal variable, with standard deviation σ, that accounts the shadowing effect. Example of σ values: 0 dB in free space, and 5.8 dB inside a building [66].

Note that many other models have also been proposed in the literature [66], but the log-normal model is the most popular one due to its simplicity.

[5] $P[mW] = 10^{P[dBm]/10}$ and $P[dBm] = 10\log_{10}(P[mW]/1mW)$.

From the channel model in Eq. (9.4), we can derive the distribution of the RSS-based range measurements

$$\hat{d} = d \cdot 10^{\frac{X_\sigma}{10n_p}} = 10^{\log_{10}(d)+\frac{X_\sigma}{10n_p}} = e^{(\ln 10)\cdot[\log_{10}(d)+\frac{X_\sigma}{10n_p}]} = e^{(\ln d)+\frac{\ln 10}{10n_p}X_\sigma}. \tag{9.5}$$

Therefore RSS-based range measurements are distributed according to a log-normal distribution

$$\hat{d} \sim \ln\mathcal{N}(\ln d, \frac{\sigma \ln 10}{10n_p}). \tag{9.6}$$

Which yields to the unbiased range estimator from the RSS measurements [47]:

$$\hat{d} = d_0 \left(\frac{P_r(d)[mW]}{P_0(d_0)[mW]}\right)^{-1/n_p} e^{-\frac{\sigma^2}{2\eta^2 n_p^2}} \quad \text{,with } \eta = \frac{10}{\ln 10}. \tag{9.7}$$

The CRLB of an unbiased range estimator is derived in [64]

$$\sqrt{Var\{\hat{d}\}} \geq \frac{(\ln 10)\sigma}{10n_p} \cdot d. \tag{9.8}$$

Equation (9.8) shows that the ranging accuracy depends on the channel parameters n_p and σ, that are environment dependent. Also it deteriorates as the distance between the transmitter and the receiver increases. Thus, in order to maintain the estimation error of less than δd, the tracked node has to be within the range of [64]

$$r_0 = \frac{10}{\ln 10} \cdot \frac{\sigma}{n_p} \cdot \delta d. \tag{9.9}$$

Finally, the result in Eq. (9.8) is generalized by [59], in the case of N reference nodes and one tracked node in the 2D case:

$$\sqrt{Var\{\hat{d}\}} \geq \frac{(\ln 10)\sigma}{10n_p} \cdot \left(\sqrt{\frac{\sum_{i=1}^{N} d_i^{-2}}{\sum_{i=1}^{N-1}\sum_{j=i+1}^{N}\left(\frac{d_{\perp ij} d_{ij}}{d_i^2 d_j^2}\right)}}\right) \tag{9.10}$$

where d_i is the distance between reference and tracked nodes, d_{ij} is the distance between reference nodes i and j, and $d_{\perp ij}$ j is the shortest distance from the tracked node to the line segment connecting nodes i and j. This result highlights the impact of the geometric distribution of the reference nodes on the localization accuracy.

Propagation Time Distance between neighboring nodes can be estimated using propagation time measurements. Namely, Time of Arrival (ToA) methods are used when reference and tracked nodes are synchronized, or Time Difference of Arrival (TDoA) that only requires synchronization between reference nodes. Time delay measurements commonly use generalized cross-correlation or matched filter receivers [22, 35].

Time of Arrival (ToA) There are two categories of ToA-based distance measurements: *one-way* propagation time, and *round-trip* propagation time measurements (cf. Fig. 9.5). The *one-way* propagation time

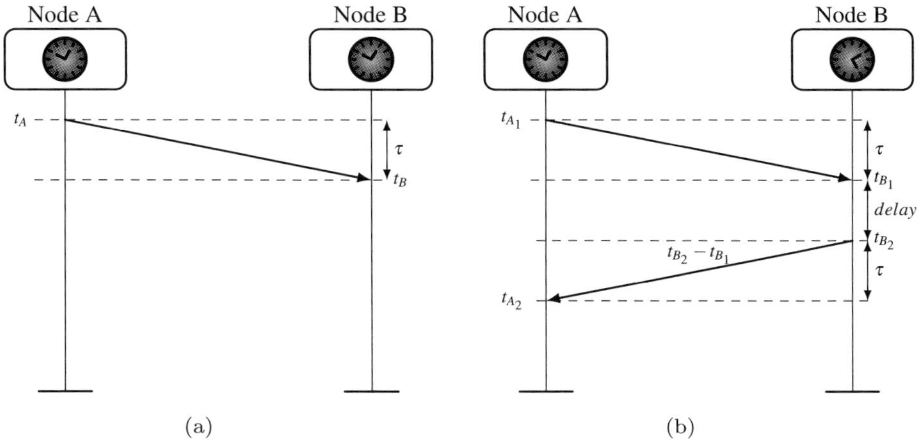

Fig. 9.5 Propagation time. (a) One way. (b) Round trip.

measurements measure the difference between sending time at transmitter node (t_A) and receiving time at receiver node (t_B). This delay is related to the inter-node distance by $\tau = d/c$. The main drawback of this technique, is that the local time of the transmitter and the receiver should be accurately synchronized. Assuming Line Of Sight (LOS) conditions and Gaussian noise at receiver level, the CRLB for one-way propagation time ranging is given by [45, 64]

$$\sqrt{Var\{\hat{d}\}} \geq \frac{c}{2\sqrt{2}\pi} \cdot \frac{1}{\beta} \cdot \frac{1}{\sqrt{SNR}}. \tag{9.11}$$

Equation (9.11) shows that, unlike RSS-based distance estimation, the accuracy can be improved by increasing the effective signal bandwidth β and/or the SNR. However, in practice, the transmitter-receiver synchronization requirement increases the cost and the complexity to the system.

In order to cope with synchronization constraint, the *round-trip* propagation time measurements measure the difference between sending time t_{A_1} at transmitter node and the time when the signal is echoed back by the receiver node t_{A_2}. That is

$$round\text{-}trip = 2 \cdot \tau = (t_{A_2} - t_{A_1}) - (t_{B_2} - t_{B_1}).$$

Note that synchronization between nodes A and B is no longer required since the same clock is used to compute the round-trip propagation time.

In practice, the internal delay for node B to echo the signal, should be considered with care in order to avoid jitter introducing meters of errors on distance measurements. For instance, a hardware implementation with *a priori* calibration is a good option.

Assuming LOS conditions, Gaussian noise at receiver level, and no changes in the bandwidth or SNR conditions, the CRLB for *round-trip* propagation time ranging is given by [45]

$$\sqrt{Var\{\hat{d}\}} \geq \frac{1}{2}\left(\frac{c}{2\sqrt{2}\pi} \cdot \frac{1}{\beta} \cdot \frac{1}{\sqrt{SNR}}\right). \quad (9.12)$$

Equation (9.12) shows that the ranging accuracy for *round-trip* propagation time is twice better than the *one-way* scenario.

With recent advances in radio technology, the UWB signals are used for accurate time-based ranging, even in challenging environments, at short distances with very low energy levels [22, 23]. Nanotron Technologies Gmbh have designed the Chirp Spread Spectrum (CSS) modulation technique [24] to cope with multi-path effects, operating in the 2.44 GHz band, with approximately 60MHz effective bandwidth. In order to avoid synchronization and eliminate clock drift and offset, Nanotron Technologies Gmbh have also developed SDS-TWR[6], an extension of the *round-trip* propagation time ranging presented earlier. The CSS modulation confers to SDS-TWR its robustness to multi-path effects. Furthermore, time-based ranging have been standardized with release of the 802.15.4a standard, that adopted both UWB and CSS signaling [29, 70].

Finally, other time-based techniques have been developed, such as the *Cricket* system [63] that combines RF communication and ultra-sound ranging. The idea behind the cricket system is to take benefit from the fact that, in the air, the speed of the sound is much smaller than the speed of the light. The radio link is only used to synchronize the ultrasound microphones. Other techniques use audible sound for time of arrival measurements such as *Beepbeep* [60] and *Beep* [46]. Finally, the *Lighthouse* approach uses laser beams to measure the distance between tiny dust nodes and the lighthouse, a modified base station device [68].

Time Difference of Arrival (TDoA) Another type of distance-based measurements is the Time Difference of Arrival (TDoA), that consists of measuring the difference between the arrival times between two signals traveling between the target and the tracked nodes. For illustrating the TDoA principle, consider the case of one tracked node and two reference

[6] Symetrical Double-Sided Two Way Ranging

nodes. In this case, the position of the tracked node is on a hyperbola, with foci the reference nodes, as illustrated in Fig. 9.6. With more than two reference nodes, the estimated position is at the intersection of the hyperbolas whose foci are at the locations of the reference nodes pairs.

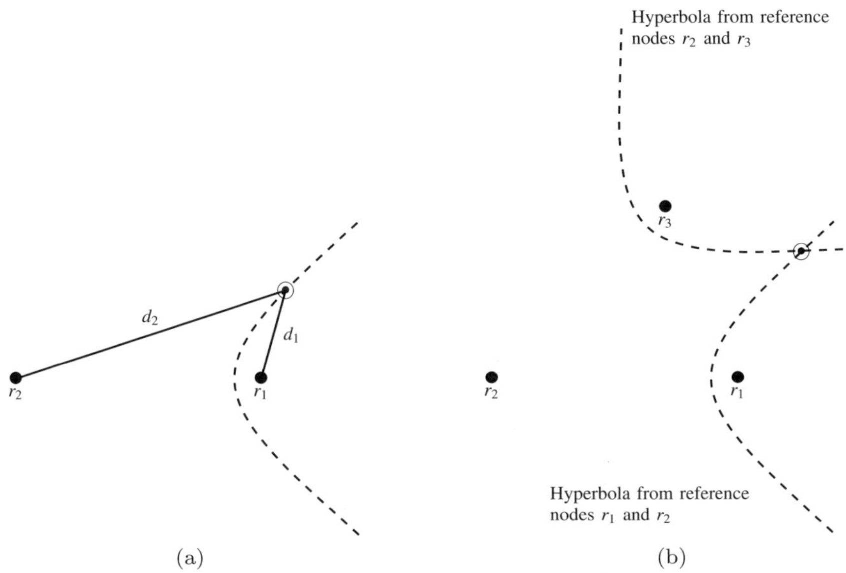

Fig. 9.6 TDOA measurements. (a) With two reference nodes. (b) With three reference nodes.

One approach for measuring TDoA is to first measure the ToA for each signal between each (reference node, tracked node) pairs as follows

$$\begin{cases} \hat{\tau}_1 = \tau_1 + \text{offset}_1 & (9.13) \\ \hat{\tau}_2 = \tau_2 + \text{offset}_2. & (9.14) \end{cases}$$

Since the reference nodes are synchronized, $\text{offset}_1 = \text{offset}_2$ which leads to the TDoA estimate

$$\hat{\tau}_{TDoA} = \hat{\tau}_1 - \hat{\tau}_2 = \frac{d_1 - d_2}{c}. \quad (9.15)$$

The corresponding CRLB expression can be deduced from the ToA case, which shows that its accuracy increases with effective bandwidth and/or SNR [23]. It is also proved in [64] that TDoA approach cannot perform better than ToA approach in terms of accuracy limits.

Finally, another approach for measuring TDoA is the generalized cross correlation method, given by

$$\hat{\tau}_{TDoA} = arg \max_{\tau} \left| \int_0^T r_1(t) r_2(t+\tau) dt \right| \tag{9.16}$$

where $r_i(t)$ is the signal between the tracked node and the ith reference node, and T the observation interval. Refer to [22] and [47] for more details and discussions.

Phase Difference Measurements The position of the tracked node can also be estimated using phase difference measurements. The below presented techniques exploit fundamental laws of physics to determine ranging information. Namely, Near field electromagnetic ranging (NFER) exploits near field phase behavior discovered by Heinrich Hertz [75], and Radio Interferometric Measurements (RIM) is inspired by the interferometric positioning in the optical regime developed thirty years earlier [14, 48].

Near Field Electromagnetic Ranging (NFER) Near field electromagnetic ranging (NFER) uses the near field phase relationship of electric and magnetic components to infer range estimate [74, 76].

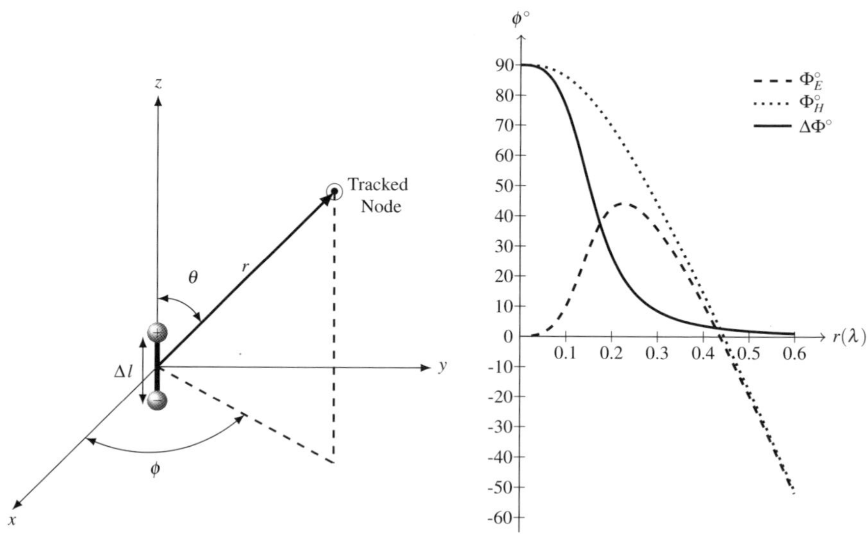

Fig. 9.7 Near field electromagnetic ranging (NFER). (a) Spherical coordinates, Hertzian dipole. (b) Phase difference versus range.

Consider a Hertzian dipole of length Δl, carrying a uniform current, $I = I_0 \cos(\omega t)$ as in Fig. 9.7. In spherical coordinates (r, θ, ϕ), the electric field component E_θ and magnetic field component H_ϕ are given by [32, 75]

$$\begin{cases} E_\theta = \dfrac{I_0 \Delta l \sin\theta}{4\pi\omega\epsilon_0}\left[\dfrac{\beta}{r^2}\cos(\omega t - \beta r) + \left(\dfrac{1}{r^3} - \dfrac{\beta^2}{r}\right)\sin(\omega t - \beta r)\right] & (9.17) \\ \\ H_\phi = \dfrac{I_0 \Delta l \sin\theta}{4\pi\omega}\left[\dfrac{1}{r^2}\cos(\omega t - \beta r) - \dfrac{\beta}{r}\sin(\omega t - \beta r)\right] & (9.18) \end{cases}$$

where $\beta = \frac{\omega}{c} = \frac{2\pi}{\lambda}$, c is the speed of light, λ is the wavelength, and ϵ is the permittivity of free space. From Eq. (9.17) and Eq. (9.18), we can compute the phase difference between E_θ and H_ϕ, which leads to [32, 75]

$$\Delta\phi = \Phi_E - \Phi_H = \cot^{-1}\left(\frac{\omega\, r}{c} - \frac{c}{\omega\, r}\right) - \cot^{-1}\frac{\omega\, r}{c}. \tag{9.19}$$

The plot of $\Delta\phi$ in Fig. 9.7 clearly shows a direct mapping between $\Delta\phi$ and the range r. By using the relationship $\cot(a - b) = -\frac{1+\cot\, a \cot\, b}{\cot\, a - \cot\, b}$, the range can be related to the phase by [76]:

$$r = \frac{\lambda}{2\pi}\sqrt[3]{\cot\Delta\phi}. \tag{9.20}$$

The CRLB of NFER is derived in [32] for both ranging and 2D-localization. For the ranging case, the CRLB is given by

$$\sqrt{Var\{\hat{d}\}} = b\left(\frac{c^6 + d^6\omega^6}{3\, c^3 d^2 \omega^3}\right) \tag{9.21}$$

with:

$$b^2 = \frac{U^2}{U^3 + V^2(1-U)},\ U = \frac{SNR_E + SNR_H}{2},\ \text{and}\ V = \frac{SNR_E - SNR_H}{2}.$$

Equation (9.21) shows that accuracy is dependent on real location, the used frequency, and the signal-to-noise ratio. In the 2D case, the CRLB additionally shows the impact of the geometrical conditioning, and that the optimal accuracy cannot be achieved by using only one frequency. Kim *et al.* have presented a scheme using multiple frequencies in order to improve accuracy [31]. In the case of the NFER, CRLB as derived in [32] becomes a tool for choosing the optimal frequency for a given coverage area.

Radio Interferometric Measurements (RIM) Radio Interferometric Ranging, exploits radio frequency interference of two waves emitted from two reference nodes at slightly different frequencies, in order to obtain the ranging information for localization by measuring the relative phase offset. In [48] authors describe the Radio-Interferometric Positioning (RIPS) localization method and provide a mean for measuring the phase difference indirectly using RSSI. In order to illustrate the principle of RIPS, consider three reference nodes A,B,C and one tracked node D as in Fig. 9.8.

Let A and B transmit at the same time, two unmodulated sine waves at two close frequencies f_a,f_b. The resulting interference is a beat signal with a beat frequency $|f_a - f_b|$. The reference node C and the tracked node D, acting as receivers, will receive the beat signal with a phase difference depending on the distance difference between the quartet (A, B, C, D)

$$\Delta\Phi = \frac{2\pi}{\lambda}(d_{AD} - d_{BD} - d_{BC} - d_{AC}) \bmod 2\pi \tag{9.22}$$

where $\lambda = \frac{c}{(f_a+f_b)/2}$, and d_{XY} is the euclidean distance between nodes X and Y. Which yields to the definition of the *q-range* measurement, defined as

$$q_{ABCD} = d_{AD} - d_{BD} - d_{BC} - d_{AC}. \tag{9.23}$$

Note that a single RIPS measurement given by Eq. (9.23), places the tracked point on a hyperbola branch (cf. Fig. 9.9) whose foci are the transmitter pair (A, B). The relative phase offset in Eq. (9.22) can then be rearranged as follows [37]

$$\Delta\phi = \frac{\Delta\Phi}{2\pi}\lambda = q_{ABCD} \bmod \lambda. \tag{9.24}$$

Note that several measurements are required at different transmit frequency pairs (f_a, f_b) in order to resolve the phase ambiguity (due to mod 2π), therefore *the q-range* can be found by solving the following system of n equations

$$\Delta\phi_i = q_{ABCD} \bmod \lambda_i. \tag{9.25}$$

In order to find the position of the tracked node, consider the case where the transmitter pair is (A, C) and the receiver pair (B, D). The associated *q-range* is

$$q_{ACBD} = d_{AD} - d_{CD} - d_{BC} - d_{AB}. \tag{9.26}$$

The position of the tracked node can be obtained by solving the system of equations obtained from Eq. (9.26) and Eq. (9.26). Geometrically, this corresponds to the intersections of the two ranging hyperbolas.

RIPS trilateration is discussed in detail in [37]. For a complete review of the interferometry in WSN refer to [71].

In [84] authors present an analytical study of the impact of RIPS measurement noise on the localization error. It appears that the localization error is small if the tracked node is located inside the triangle $\triangle ABC$, and generally increases at a steady exponential rate as the tracked node moves away from the triangle, unless it is close to L_{ABC}. Where L_{ABC} is the union of the three lines representing the extensions of the sides of the triangle $\triangle ABC$, but not including the interiors of the edges of $\triangle ABC$.

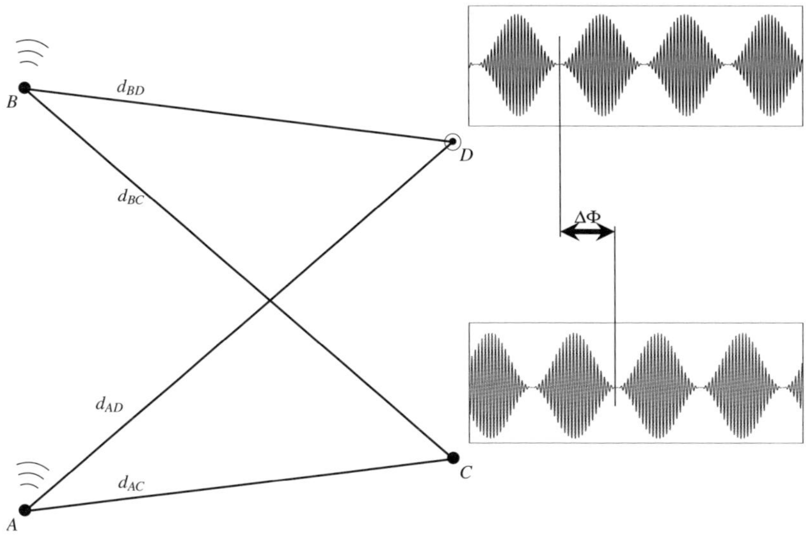

Fig. 9.8 RIPS measurement technique.

9.2.2 *Network related measurements*

Network connectivity measurements are possibly the simplest measurements. The tracked node location can be inferred by analyzing its neighboring reference nodes in terms of connectivity, radio coverage area, and neighborhood proximity [43]. This kind of measurements are very cost effective and straightforward in large-scale networks.

In connectivity measurements, a node measures the number of nodes in its transmission range. This measurement defines a proximity constraint between these two nodes, which can be exploited for localization [9, 15]. For

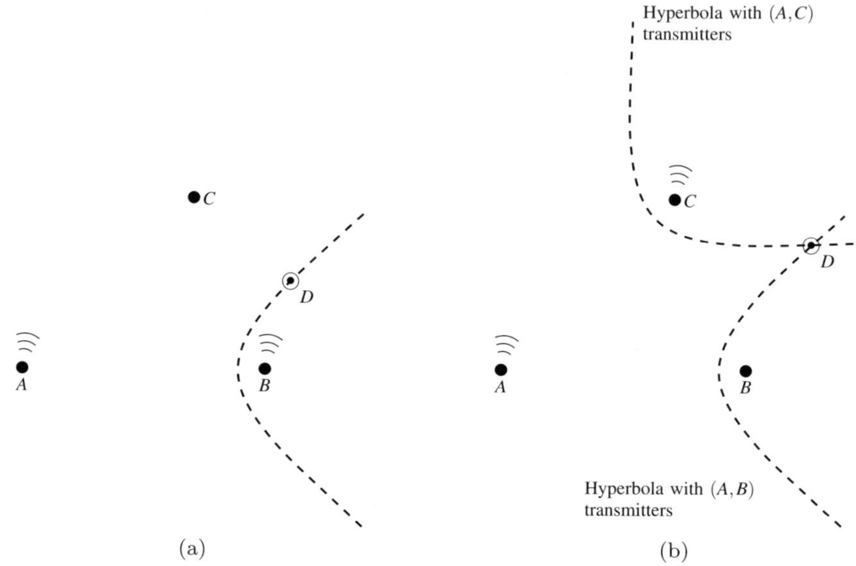

Fig. 9.9 RIPS localization rounds. (a) First Round. (b) Second Round.

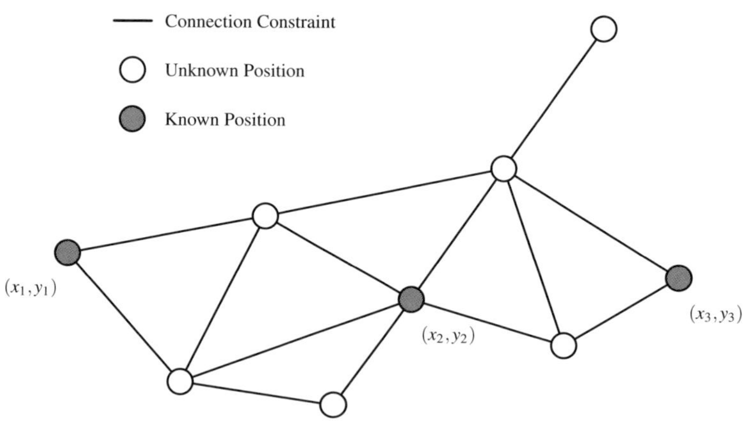

Fig. 9.10 Graph illustrating connectivity constraints.

instance, when a tracked node detects three neighboring reference nodes, it can assume to be close to these nodes and estimate its location as the centroid of the three reference nodes [26].

9.3 Localization Theory and Algorithms

In this section, we give a brief introduction to some fundamental theories in sensor network localization, and a set of major sensor network localization algorithms are discussed.

The objective behind a positioning methodology is to determinate the *location information* of a number of nodes. Location information can be interpreted as any form of location indicator such as exact location, the deployment area or the location distribution. As we have seen in Sec. 9.2, parameters extracted from signals traveling between the nodes will allow to establish pairwise spatial relationships (angle, distance or proximity). Localization algorithms will take those parameters as inputs to estimate the position of the target nodes according to a certain strategy.

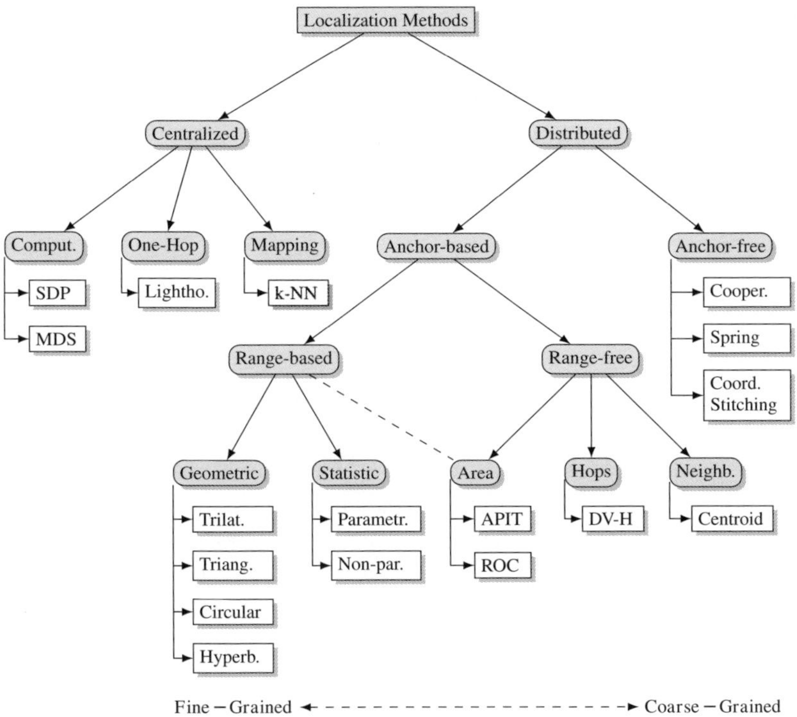

Fig. 9.11 Location methods overview.

There exist well organized surveys in the literature that propose different classifications of the localization systems based on different criteria at different hierarchical levels such as the number of hops, the presence of reference nodes (anchors) or the computational organization [4, 18, 20, 47, 56, 58, 73, 83]. The purpose of this section is not to provide definitive and exhaustive taxonomy of the different localization methods. Instead, this section is intended to present a comprehensive introduction to some of today's most popular localization methods. A special effort has been made to conciliate approaches from different authors to render a general schema. In the Fig. 9.11, a set of the most representative localization methods are displayed within some of the implementation choices associated to them.

We have first categorized the localization algorithms into two main types; namely centralized and distributed, based on the direct dependency on a centralized resource.

9.3.1 *Centralized methods*

Centralized localization methods present a direct dependency on a centralized resource. This resource can be some information previously collected (*mapping*), a central machine with powerful computational capabilities (*centralized computing*) or some kind of *one hop* location reference (landmark, satellite...) providing this centralized service.

9.3.1.1 *Centralized computing*

Centralized computing basically migrates inter-node ranging information and connectivity data to a sufficiently powerful central base station to be processed. All other nodes in the network only gather the location related information, such as RSS, and send it to the central processing location. The base station calculates the estimated location of all the tracked nodes and communicates it back if requested.

The advantage of centralized processing is minimizing the required capabilities (e.g., processing power and memory space) of the nodes, excepting the central location processing node. This benefit however comes with a communication cost, creating high traffic levels and increasing latency since all the nodes must communicate with a single central receiver to determine their location. The high level of traffic can cause bottlenecks in the network and limit the location update rate. The latency and traffic problems get worse increasing the size of the network. Therefore the centralized processing method is more suitable for small network or a network where the location update rate is low.

Two representative proposals in this category are *Semidefinite Programming (SDP)* techniques [15] and *Multi-Dimensional Scaling (MDS)* [78].

Semidefinite Programming (SDP) Many localization problems can be formulated as a convex optimization problem. They can be solved using *linear* and *semidefinite programming (SDP)* techniques [15]. SDP is a generalization of linear programming and has the following form:

$$\begin{array}{ll} \text{Minimize} & c^T x \\ \text{Subject to:} & F(x) = F_0 + x_1 F_1 + \ldots + x_N F_N \leq 0 \\ & Ax < B \\ & F_k = F_k^T \end{array} \tag{9.27}$$

where $x = [x_1 y_1, x_2 y_2, \ldots, x_m y_m, x_{m+1} y_{m+1}, \ldots, x_N y_N]^T$. The first m entries are fixed and correspond to the reference nodes positions and the remaining $n - m$ positions are computed by the algorithm. The objective is to find a possible position for each target node when a the position of a set of reference nodes is given. Proximity constraints imposed by known connections can be represented as *linear matrix inequalities (LMIs)*. In the case of nodes communicating within a perfect circle, the estimated region is convex and can not be described by linear equations but as an LMI. For a maximum radio range R_{max}, if two nodes, with positions x_i and x_j, are in communication, their separation must be less than R_{max}, i.e., it exists a proximity constraint between them. This can be represented as a *radial constraint* and expressed as a LMI: $\|x_j - x_i\| \leq R_{max}$.

The advantage of this method is that it is simple to model hardware that provides ranges or angles and simple connectivity. SDP simply finds the intersection of the constrains. There are efficient computational methods available for most of convex programming problems. However, this estimation methodology requires centralized computation. To solve the optimization problem, each node must report its connectivity to a central computer. This approach also requires to handle large data structures and lacks of scalability because of its complexity. The relevant operation for radial data is $O(n^3)$ where n is the number of convex constraints.

Multi-Dimensional Scaling (MDS-MAP) *Multi-dimensional Scaling (MDS)* [78] is often used as part of information visualization techniques for exploring similarities or dissimilarities. It displays the structure of distance-like data as a geometric picture.

The typical goal of MDS is to create a configuration of n points in one, two or three dimensions. Only the inter-point *distances* are known. MDS enables the reconstruction of the relative positions of the point based on the pairwise distances.

Typical procedure of MDS involves three stages:

- First, compute the shortest path between all pairs of nodes. The shortest path distances are used to construct the distance matrix for MDS. The (i, j) entry represents the distance along the shortest path from the node i to j. If only connectivity information is available, that distance will be the number of hops. However, it can also incorporate distance information between neighboring nodes when it is available.
- In a second stage, classical MDS is applied to the distance matrix to obtain estimated relative node positions.
- Finally , the relative positions are transformed to absolute positions with the help of some number of fixed anchor nodes.

The strength of MDS-MAP is that it can be used when there are few or no anchor nodes. It can use both connectivity and distance measurement ranging techniques and provides both absolute and relative positioning. However, the main problem with MDS is its poor asymptotic performance, which is $O(n3)$.

More detailed work based on MDS can be found in [1].

9.3.1.2 *One-Hop positioning*

This kind of positioning methods require *line of sight (LOS)*, i.e., direct contact, between the reference position, the landmark, and the node to locate. The most representative solution of this class is the GPS, where receiver has to have a clear line of sight to the satellites to operate.

An interesting approach to estimate distance between an optical receiver and transmitter is the *Lighthouse* [68] location system. It is a laser-based solution which allows nodes to autonomously estimate their location with high accuracy without additional infrastructure components besides a base station device. The transmitter consists on a parallel optical beam rotating at a constant speed. The receiver is equipped with an optical sensor and a clock. Measuring the time it *sees* the beam (t_{beam}), the distance (d), from the base station can be calculated when the rotational speed (or the time it takes for a complete rotation, (t_{turn}) and the width of the beam (b), are known:

$$d = \frac{b}{2\sin(\pi t_{beam}/t_{turn})}. \tag{9.28}$$

As a result we have a simple ranging system where all the potential positions of the receiver form a cylinder with radius d centered at the lighthouse rotation axis. Using three such lighthouses in different placements, the location of the node can be inferred with trilateration principles. Nevertheless, the original proposal aimed to have a unique base station. For that purpose, distances are measured in the three axes of the space using mutually perpendicular rotation axes in single positioning device. A major advantage of this system is that the optical receiver can be very small in cost and size. However the transmitter may be large and expensive and the LOS requirement remains a big handicap in many practical cases.

9.3.1.3 *Profiling techniques*

In Sec. 9.2 we have seen different ways to estimate distances between nodes. Localization algorithms can then be applied to these distances to obtain the estimated position of the tracked node. Nevertheless, wireless sensor network environments, and specially indoor environments, are often complicated to model and their model parameters determination is also a difficult task. Such a challenging scenario can be overcome using another approach, namely *profiling-based techniques* [5, 36, 61, 69].

The main idea behind these localization techniques, also referred to as *mapping* or *fingerprinting*, is to determine a regression scheme based on a set of training data and then to estimate the position of a given node according to that regression function. They work by first constructing a kind of map of the signal parameters behavior for each anchor node over a coverage area. In addition to anchor nodes, a collection of n sample points with a priori chosen positions must be defined to collect that training data. At each location, $l_i = (x_i, y_i)^T$, a vector of signal parameters m_i is obtained. Typically the m_{ij} entry corresponds to the value of the *signal strength* from/at the anchor j when the anchor node is at location l_i although other signal parameters may be used.

These *training data* can then be expressed as:

$$\tau = \{(m_1, l_1), (m_2, l_2), \ldots, (m_n, l_n)\}\,. \tag{9.29}$$

For the training set given in Eq. (9.29), a *position estimation rule* must then be determined, i.e., a pattern matching algorithm or regression function to estimate the location l of a given target node based on a parameter vector m related to the target node. Some common mapping techniques used in location estimation include *k-nearest neighbors (k-NN)* estimation, *support vector regression (SVR)* and neural networks [16, 21, 39, 49, 55]. As in [20] we develop *k-NN* in order to provide an intuition of a simple mapping-based position estimation.

k-nearest neighbors (k-NN) In its simplest version *1-NN* determines the estimated position of a target node at the location l_j in the training set τ that has the associated vector m_j with the shortest Euclidean distance to the measured parameter vector m:

$$j = \arg_{i \in 1,\dots,n} \min \parallel m - m_i \parallel . \tag{9.30}$$

In general, *k-NN* determines the position of the target node with the help of the k parameters vectors in τ that have the smallest distances to the given parameter vector m. The position $\hat{l}$ is then estimated by the weighted sum of the positions corresponding to those k nearest parameter vectors:

$$\hat{l} = \sum_{i=1}^{k} w_i(m) l_i, \tag{9.31}$$

where w_i is the weighting factor associated to the ith reference location. Various weights can be used as studied in [49]. For instance, in the uniform weighted scheme, the sample mean of position is used:

$$\hat{l} = \frac{1}{k} \sum_{i=1}^{k} l_i. \tag{9.32}$$

The main advantage of mapping techniques is that they have a certain degree of inherent robustness. They can provide very accurate position estimation in challenging environments with multipath and non-line of sight propagation. However, the main disadvantage is the requirement that the training database should be large enough and representative of the current environment for accurate position estimation. The database should be updated frequently enough so that channel characteristics in the training and position estimation phases do not differ significantly. Such an update requirement can be very costly for positioning systems operation in dynamic environments, such an outdoor positioning system.

9.3.2 *Distributed algorithms*

One way to overcome the traffic bottleneck of the centralized processing method is to divide the network into sections and allocate a node capable of executing the positioning algorithm to each section. An alternative approach, is to distribute the location-estimation task among almost all the nodes in the network. In this way, there is no centralized location processing node and each node determines its own location by communication only with nearby anchor nodes and potentially other tracked nodes. In a fully distributed processing method, all the nodes must satisfy certain processing capabilities and memory space requirements. One of the advantages of distributed processing is relatively uniform packet traffic, which makes it easy to expand the traffic of the network.

Distributed algorithms can be classified according to whether they use pre-configured reference positions (anchor-based vs anchor-free) or the granularity of the measured employed, i.e., whether they make some kind of range (distance or angle) correlation (range-based vs range-free).

9.3.2.1 *Anchor-based techniques*

Anchor-based algorithms assume that a certain number of nodes, referred to as *anchors* or *beacons*, know their own position through manual configuration or an external positioning system such as GPS. Tracked nodes' location is then determined by referring to that reference positions with the help of inter-sensor measurements such us the ones we have seen in Sec. 9.2.

Depending on the measurement techniques employed, anchor-based algorithms can be classified [43] from fine-grained to coarse-grained into several categories such as: *location*, *distance*, *angle*, *area*, *hop-count* and *neighborhood* (see Fig. 9.11). This classification allows us to broadly distinguish localization algorithms between *range-based* and *range-free* [26]. Range-based approaches rely on signal features such as signal strength, time of flight or angle of arrival for calculating relative distances or angles. In contrast, range-free methods do not try to estimate direct point-to-point distance from the received signal parameters; they use topological information (connectivity, signal comparison ...) rather than ranging.

Range-based localization techniques Range-based localization applies geometric techniques to estimate the position of a target node. It uses a set of position related parameters in a number of reference nodes to describe geometric figures that are supposed to intersect in the point of interest.

Geometric methods As we have seen in Sec. 9.2, some measurements parameters can define a geometric figure of uncertainly around the anchor node. The position of the target node can be estimated as the intersection of those figures. The received signal strength or the time of arrival of the signal determine a circumference by translating that parameter to physical distance. Using *trilateration* the estimated location for the target node is given by the intersection of three circles from three different references, (x_i, y_i) (see Fig. 9.12(a)). In the case of the time difference of arrival, the geometric figure corresponds to an hyperbola and an intersection point can also result as an estimation. On the other hand, the angle of arrival measure defines a straight line passing through both, the target and the reference nodes In this case, two parameters are enough to calculate the estimated position via *triangulation*.

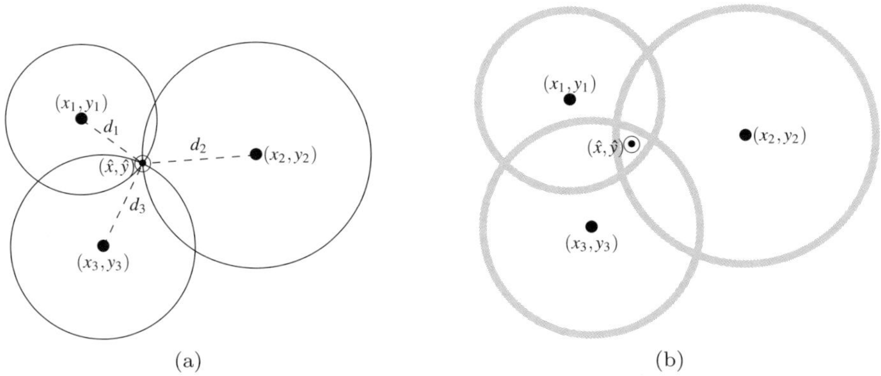

Fig. 9.12 Location estimation using Trilateration. (a) Ideal case. (b) Including range error.

Unfortunately, in a practical implementation, the ranging measurements contain noises. The presence of fading and shadowing may lead this method to produce no results at all. Error in distance estimation can prevent the bearing lines to have a common intersection point (see Fig. 9.12(b)). Thus, an optimization algorithm must be applied to choose an estimated location according to some criteria. Two sample algorithms are the circular and the hyperbolic positioning algorithms. The former optimizes directly the error associated to distances while the latter makes distance-difference optimization.

Circular positioning algorithm The *Circular Positioning* algorithm [40] adopts the criterion of minimizing the sum square error ε. This error can be expressed as:

$$\varepsilon = \sum_{i=1}^{n} \left(\sqrt{(x_i - x)^2 + (y_i - y)^2} - d_i \right)^2, \tag{9.33}$$

where (x_i, y_i) is the position of each reference node. The position (x, y) of the node that minimizes that error can be calculated using the steepest descend method defined by:

$$\begin{bmatrix} \hat{x} \\ \hat{y} \end{bmatrix}_{k+1} = \begin{bmatrix} \hat{x} \\ \hat{y} \end{bmatrix}_{k} - \alpha \begin{bmatrix} \frac{\delta \varepsilon}{\delta x} \\ \frac{\delta \varepsilon}{\delta y} \end{bmatrix}_{x=\hat{x}_k, y=\hat{y}_k} . \tag{9.34}$$

This method requires an initial location to begin the iteration, which can be the midpoint of the reference positions under consideration.

Hyperbolic positioning algorithm The *Hyperbolic Positioning* algorithm [40] also referred to as *Linear Least Squared errors (LLS)* [18] does not minimize directly the sum of the squared errors of the erroneous distance estimations to reference positions as in the previous case. Instead it minimizes a linear function of it by subtracting two distances estimations i.e., it minimizes the sum of the distance to the hyperbolas resulting from the subtraction.

Considering n reference positions we can write the distance estimations $(d_{i=1\ldots n})$ to the target node as:

$$d_i^2 = (x - x_i)^2 + (y - y_i)^2. \tag{9.35}$$

To solve the previous system of equations a linearization is performed by subtracting the location of the first reference from all other equations [54]. The resulting system of equations can be expressed in the form $Ax = b$ as:

$$\begin{bmatrix} 2x_1 - 2x_2 & 2y_1 - 2y_2 \\ 2x_1 - 2x_3 & 2y_1 - 2y_3 \\ \cdots & \cdots \\ 2x_1 - 2x_n & 2y_1 - 2y_n \end{bmatrix} \begin{bmatrix} x \\ y \end{bmatrix} = \begin{bmatrix} d_2^2 - d_1^2 + x_1^2 - x_2^2 + y_1^2 - y_2^2 \\ d_3^2 - d_1^2 + x_1^2 - x_3^2 + y_1^2 - y_3^2 \\ \cdots \\ d_n^2 - d_1^2 + x_1^2 - x_n^2 + y_1^2 - y_n^2 \end{bmatrix} . \tag{9.36}$$

Therefore, the estimated position of the target node can be calculated as the least squares solution of this equation given by:

$$\begin{bmatrix} \hat{x} \\ \hat{y} \end{bmatrix} = (A^T A)^{-1} (A^T b). \tag{9.37}$$

Both, circular and hyperbolic algorithms, give the same weight to the different distance estimations. Nevertheless, as we have seen in Sec. 9.2, measurements such as received signal strength (RSS) do not depend linearly on the distance between the nodes. From Eq. (9.4) it can be deduced that the same error in the RSS measurement will produce larger errors in the distance estimation if the distance between the nodes is higher. That is, the accuracy of the distance estimations depends on the distance itself. The use of weighted techniques to improve the accuracy of the hyperbolic and circular positioning algorithms respectively has been proposed [80]. They give more weight to those measurements corresponding to short distances, which accuracy is expected to be greater.

Statistical location techniques Unlike the geometric techniques, the statistical approach presents a theoretical framework for position estimation for multiple measurement parameters with or without the presence or noise.

In order to formulate this generic framework, consider the following model for each of the N estimated parameters:

$$z_i = f_i(x, y) + \eta_i, \tag{9.38}$$

where η_i is the noise at the corresponding estimation and $f_i(x, y)$ is the real value of the signal parameter at the position (x, y).

As we have seen in Sec. 9.2 for ToA/RSS, AoA and TDoA, $f_i(x, y)$ can be expressed as:

$$f_i(x, y) = \begin{cases} \sqrt{(x - x_i)^2 + (y - y_i)^2} & \text{ToA/RSS} \\ \tan^{-1}\left(\frac{(y - y_i)}{(x - x_i)}\right) & \text{AoA} \\ \sqrt{(x - x_i)^2 + (y - y_i)^2} - \sqrt{(x - x_0)^2 + (y - y_0)^2} & \text{TDoA} \end{cases} \tag{9.39}$$

In the case that the probability density function of the noise η is known for a set of parameters, *parametric* approaches such as *Bayesian* and *Maximum Likelihood (ML)* can be used. Those techniques are studied in detail in [20]. In the absence of that information *non-parametric* methods must be used. Actually, *profiling techniques*, such as *k-NN*, *SVR* and *neural networks* approaches referred in Sec. 9.3.1.3, are examples of non-parametric estimators since they do not make any assumption concerning the density of probability function of the noise.

Range-free localization techniques Cost and hardware limitations in wireless sensor nodes often prevent the use of range-based localization schemes. For some applications coarse accuracy is sufficient and range-free solutions have been revealed as a valid cost-effective alternative. Range-free based solutions do not try to estimate absolute distances among nodes using any signal feature such as signal strength, angle of arrival or time or flight. They, alternatively, use coverage range, i.e., connectivity, or comparative features between signals. Three approaches can be distinguished according to the *granurality* of the measurement employed: area, hop-count and neighborhood based.

Area based Signals coming from beacon nodes can define coverage areas described by geometric shapes. The area based location estimation method will compute the intersection of those coverage areas and will give the centroid of this region as the resulting location estimation for the tracked node. For instance, if a tracked node receives a signal from another anchor node, a circular region, centered in that anchor node and radius its maximum coverage distance, is delimited. When several reference nodes can be listened, the overlapped area of those circles will determine an estimated location for the tracked node (see Fig. 9.13(a)). This can be extended to other scenarios. For example when angular sector can be determined for the incoming signal from the beacon nodes or when lower coverage bounds are also available to describe different geometric figures (see Fig. 9.13(b)). More detailed work about localization based on connectivity-induced constraints can be found in [15].

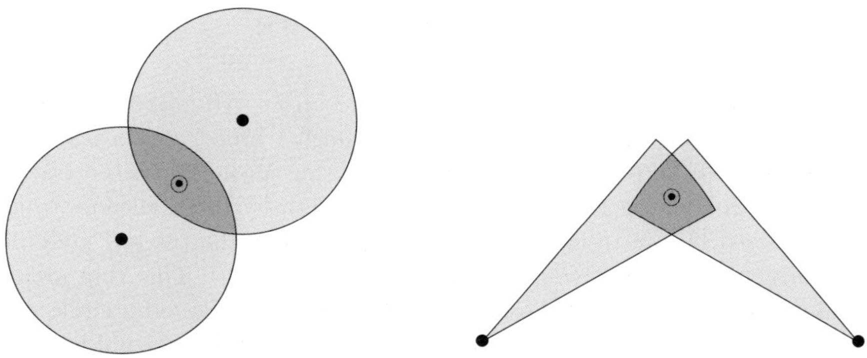

(a) (b)

Fig. 9.13 Area measurements. (a) Circles overlapping. (b) Sectors overlapping.

One popular area-based range-free location estimation scheme is *APIT* [26]. The *APIT* algorithm isolates the environment into triangles (see Fig. 9.14). The vertexes of these triangular regions are anchors nodes that the tracked node can hear. The presence inside or outside those triangular regions allows to narrow down the area in which the tracked could reside. The estimated position is obtained from the centroid of the area provided by the intersection of the reference triangles that contains the tracked node.

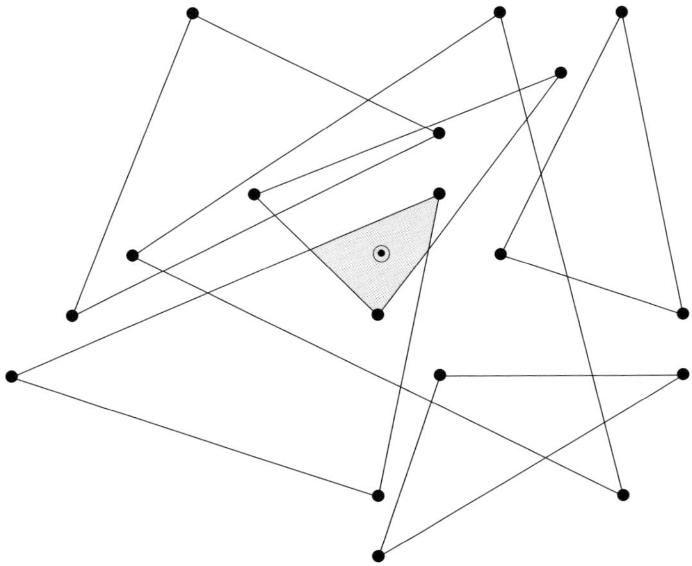

Fig. 9.14 APIT algorithm overview.

Another interesting approach is the *Ring Overlapping Circles (ROC)* [42] algorithm (see Fig. 9.15). Each anchor broadcasts beacon messages that will be received by both, neighboring anchors and the tracked node. By comparing the received signal strength by those anchors to the one received by the tracked node, the region where the tracked node lies within can be determined (in light gray in Fig. 9.15). This ring area is centered in the beacon anchor node and has as higher bound a circle with radius equal to the distance to the anchor which received signal strength is immediately inferior to the one received by the tracked node. The lower bound of the ring is delimited by a circle with radius equal to the distance to the anchor node with received signal strength immediately superior. The process is repeated by each anchor node resulting in several overlapping rings. Finally, the center of gravity of the overlapped area (in dark gray in Fig. 9.15) is reported as estimated position.

Two key assumptions are made by the ROC algorithm. Firstly, the received signal strength decreases monotonically with the distance, so we can conclude that a node that receives a higher signal strength is closer. Secondly, the antennas are supposed to be isotropic. Nevertheless, the algorithm is proclaimed to be resilient to irregular radio propagation patterns

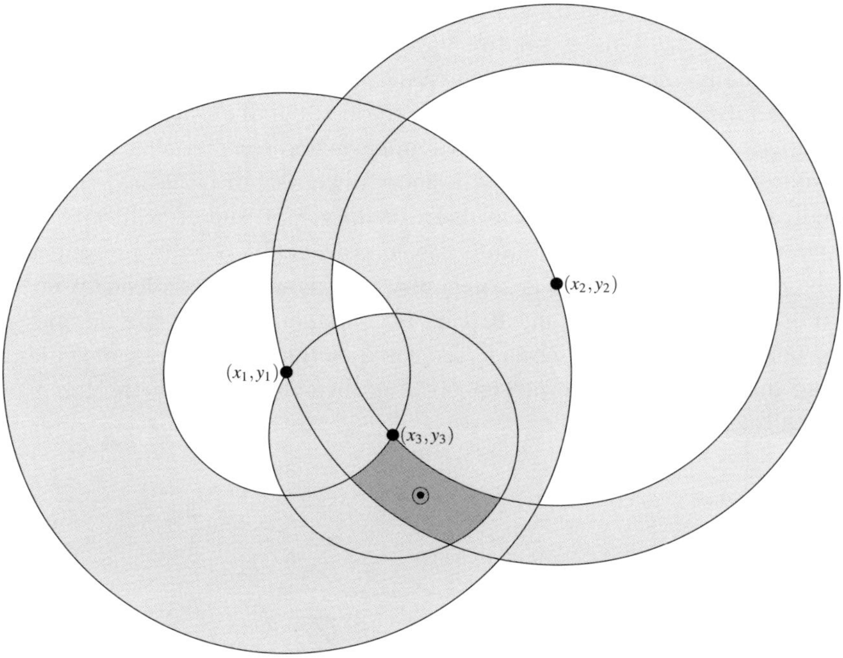

Fig. 9.15 Ring Overlapping Circles algorithm with 3 anchors.

and capable to achieve better performance than APIT with less communication overhead [41].

Hop counter If the maximum radio range among nodes is well-known, their distance from each other can be determined to be inferior to that range with high probability. *DV-HOP* [57] algorithm uses this connectivity measurements to determine the location of a node. All the anchor nodes will broadcast a beacon message that will be propagated through the network. This message includes the anchor node location and a *hop-counter* that will be incremented at every hop. Each anchor node keeps the minimum hop-counter value per anchor. This procedure enables all the nodes in the network (including anchors) to get the shortest distance (least number of hops) to anchors. To translate hop-count to physical distance, an anchor i with position (x_i, y_i) estimates the average single hop distance h_i with the following formula:

$$h_i = \frac{\sum \sqrt{(x_i - x_j)^2 + (y_i - y_j)^2}}{h_{ij}}, \tag{9.40}$$

where h_{ij} is the minimum number of hops to another anchor node j with position (x_j, y_j). This estimated hop size is then propagated to nearby nodes. Finally, once the distance estimation is made to at least three anchors, triangulation is used to report the estimated position. The main advantages of this algorithm are its simplicity and the fact that it does not depend on measurement error. The more anchors can be heard, the more precise the localization is. The main drawback is that it will only work for isotropic networks. When an obstacle prevents an edge from appearing in the connectivity graph the hop-counter methodology can lead to an inaccurate location estimation. In Fig. 9.16 we can see how the number of hops between node A and node C are equal to the hop count between node B and node D due to the presence of an obstacle, although the later are physically closer.

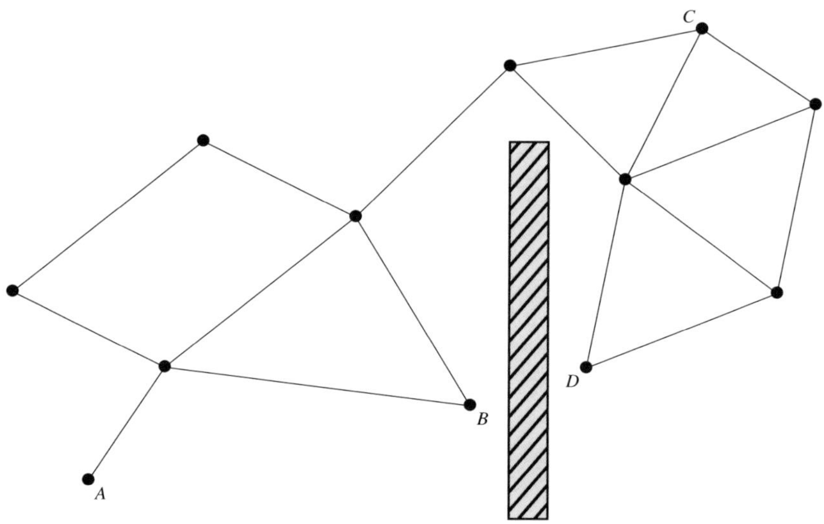

Fig. 9.16 Hop-counter with obstacle, example.

The *DV-Distance* algorithm is presented together with DV-hop proposing a similar method but distances between neighboring nodes are used instead of hops. Many other modifications of this algorithm to improve performance under certain network conditions can be found in literature [65, 77].

The *Amorphous* algorithm [53] proposes a different approach to DV-Hop to calculate the average single hop distance. It uses the density of the network, n_{local}, to correct the average hop distance estimation, d_{hop}, with the help of the Kleinrock and Silvester formula [34] for a maximum radio range R:

$$d_{hop} = R\left(1 + e^{-n_{local}} - \int_{-1}^{1} e^{\frac{-n_{local}}{\pi}(\arccos t - t\sqrt{1-t^2})} dt\right). \tag{9.41}$$

Neighborhood measurements One of the simplest coarse-grained localization methods is using the connectivity measurement, which is more robust to unpredictable environments, for neighbor proximity. The only decision to make is whether a node is within the range of another. Reference nodes can be deployed through the localization area determining non-overlapping regions. When a tracked node receives a beacon from an anchor, it will consider that reference position as its own position.

In the case of anchors (reference positions) with overlapping regions of coverage, *Centroid Location (CL)* [9] can be used. The tracked node can listen to a given subset of anchor beacons containing their reference positions (x_i, y_i) to infer its proximity to them. The node will calculate its estimated position using the following centroid formula:

$$(\hat{x}, \hat{y}) = \left(\frac{x_1 + ... + x_N}{N}, \frac{y_1 + ... + y_N}{N}\right). \tag{9.42}$$

The same authors have also proposed a reduction of the estimation error placing additional anchors using a novel density adaptive algorithm, *HEAP* [10].

Another way to ensure a localization improvement is including weights when averaging the coordinates of the beacon nodes. This is the *Weighted Centroid Location (WCL)* [8] algorithm.

The weight is a function depending on the distance and the environment conditions so different weights may be used. Small distances to neighboring anchors lead to a higher weight than to remote anchors. To calculate the approximated position of a tracked node i, every reference location j, from the n anchor nodes in range, obtains a weight w_{ij} depending on the distance:

$$(\hat{x_i}, \hat{y_i}) = \frac{\sum_{j=1}^{n} (w_{ij} \cdot (x_j, y_j))}{\sum_{j=1}^{n} w_{ij}}. \tag{9.43}$$

To determine the associated weight to a reference either the *link quality indication (LQI)* or *Received Signal Strength indicator (RSSI)* could be used [7]. Nevertheless, in the LQI case, if all the references in range provide relative high values the influence of one anchor's LQI becomes relative low. The *Adaptative WCL (AWCL)* [6] algorithm proposes to compensate high

LQI values giving more influence to the differences between the LQIs instead of the nominal values. It reduces measured LQI values of each reference in range by a part q of the lowest LQI (Eq. (9.44)),

$$(\hat{x_i}, \hat{y_i}) = \frac{\sum_{j=1}^{n}((LQI_{ij} - q \cdot min(LQI_{1...n}) \cdot (x_j, y_j))}{\sum_{j=1}^{n}(LQI_{ij} - q \cdot min(LQI_{1...n}))}. \tag{9.44}$$

A *Selective Adaptive Weighed Centroid Localization (ASWCL)* [19] approach has also been proposed to improve the accuracy by adapting the weights according to their statistical distribution.

9.3.2.2 *Anchor-free techniques*

Anchor-based algorithms have some limitations because they need another positioning scheme to place the beacon nodes. In some cases, the environment may prevent the use of such positioning system (e.g., GPS and indoor locations) so pre-configured anchors providing known reference positions are not available. In addition, the practice reveals that a large number of beacons must be deployed to provide an acceptable positioning error [11]. They require a deployment effort and they may not scale well. In contrast, anchor-free algorithms are able to determine each node relative coordinates using local distance information and without relying on beacons that are aware of their positions. Note that no absolute positions are obtained, but this is a fundamental limitation of the problem statement and not part of the algorithm itself. The relative coordinate space should be able to be translated to any other global coordinate system easily. The centralized MDS algorithm (See 9.3.1.1) is a sample of anchor-free algorithm that can obtain final absolute positions with the help of an additional step involving three or more beacons. Some popular distributed anchor-free approaches are *relaxation-based* algorithms and *coordinates stitching*.

Relaxation-based algorithms These approaches are coarse grained localization methods with a refinement phase where typically each node corrects its position to optimize a local error metric. We will briefly introduce two of the most popular relaxation-based approaches.

Cooperative ranging In the *cooperative ranging* methodologies, every single node plays the same role, and repeatedly and concurrently executes the following functions:

- Receive ranging and location information from neighboring nodes.
- Solve a local localization problem.

- Transmit the obtained results to the neighboring nodes.

After some repetitive iterations the system will converge to a global solution.

The local localization problem is revolved by making assumptions when necessary and compensating the error through corrections and redundant calculations as more information becomes available. These assumptions are needed at first in order to deal with the under-determined set of equations presented by the first few nodes. The *Assumption Based Coordinate (ABC)* algorithm [72] propose the following procedure from the perspective of a node n_0:

- The node n_0 is located at the position $(0, 0, 0)$.
- The fist node to establish communication, n_1, is placed at $(r_{10}, 0, 0)$ where r_{10} is the estimated distance from some signal parameter.
- The location of the next node n_2, (x_2, y_2, z_2), is determined using the estimated distance to both n_0 and n_1 and assuming that $y_2 > 0$ and $z_2 = 0$,

$$\begin{aligned} x_2 &= \tfrac{r_{01}^2 + r_{02}^2 + r_{12}^2}{2r_{01}} \\ y_2 &= \sqrt{r_{02}^2 + x_2^2} \\ z_2 &= 0. \end{aligned} \tag{9.45}$$

- Next location n_3 (x_3, y_3, z_3) is obtained with the only assumption that the square involved in finding z_3 is positive,

$$\begin{aligned} x_3 &= \tfrac{r_{01}^2 + r_{03}^2 + r_{13}^2}{2r_{01}} \\ y_3 &= \tfrac{r_{03}^2 - r_{23}^2 + x_2^2 + y_2^2 - 2x_2x_3}{2y_2} \\ z_3 &= \sqrt{r_{03}^2 + x_3^2 + y_3^2}\,. \end{aligned} \tag{9.46}$$

From this point forth, the system of equations used to solve for further nodes is no longer under-determined, and so the standard algorithm can be employed for each new node. Under ideal conditions, this algorithm thus far will produce a topologically correct map with a random orientation relative to a global coordinate system.

The main advantage of this approach is that global resources for a centralized computing are not required. Nevertheless, the convergence of the system may take some time and nodes with high mobility may be hard to cover.

Spring Model The *AFL (Anchor-Free Localization)* [62] algorithm, also referred to as *Spring Model*, describes a fully decentralized algorithm where nodes start from a random initial coordinate assignment and converge to a consistent solution using only local node interactions. The key idea in AFL is fold-freedom, where nodes first configure into a topology that resembles a scaled and unfolded version of the true configuration, and then run a force-based relaxation procedure.

The AFL algorithm proceeds in two phases:

- The first phase is a heuristic that produces a fold-free graph embedding which *looks similar* to the original embedding. Five reference nodes are chosen, one in the center n_0, and four in the periphery, n_1, n_2, n_3 and n_4, where the couples (n_1, n_2) and (n_3, n_4) are roughly perpendicular to each other. The choice of these nodes is performed using a hop-count approximation to distance (e.g., the first peripheral node is selected maximizing the number of hops to the initial node, $\max h_{0,1}$). Finally a node n_5 is selected and supposed to be centered by minimizing the distance in hops between n_1 and n_2 ($\min |h_{1,5} - h_{2,5}|$) and the distance between n_3 and n_4 ($\min |h_{3,5} - h_{4,5}|$) for contender nodes. Now, for all nodes n_i, the heuristics approximate the polar coordinates using the maximum radio range, R, as follows:
$$\begin{aligned} \rho_i &= h_{i,5}R \\ \theta_i &= tan^{-1}\left[\frac{(h_{1,i}-h_{2,i})}{(h_{3,i}-h_{4,i})}\right]. \end{aligned} \tag{9.47}$$
- The second phase uses a mass-spring based optimization to correct and balance localized errors. It runs concurrently at each node. At any time any node n_i has a current estimated position $\hat{p_i}$ that periodically sends to its neighbors. Using these positions, the distance $\hat{d}_{ij}$ to each neighbor n_j is estimated. Also knowing the *measured* distance r_{ij} to n_j, a force $\vec{F}_{ij}$ in the direction $\vec{v}_{ij}$ (unit vector from $\hat{p}_i$ to $\hat{p}_j$), is given by Eq. (9.48),
$$\vec{F}_{ij} = \vec{v}_{ij}(\hat{d}_{ij} - r_{ij}). \tag{9.48}$$
The resultant energy E_i of node i due to the difference of the measured and the estimated distances between nodes, can be expressed in terms of the square of the magnitude of the forces $\vec{F}_{ij}$ as Eq. (9.49),
$$E_i = \sum_j E_{ij} = \sum_j (\hat{d}_{ij} - r_{ij})^2. \tag{9.49}$$

The main advantage of relaxation based algorithms is that they are fully distributed and concurrent and they operate without anchors nodes. Nevertheless, while the computational is modest and local, it is unclear how these algorithms scale to much larger networks [4]. Furthermore, there are no provable means to avoid local minima, which could be even worse at larger scales. Traditionally, local minima have been avoided by starting the optimization process at a favorable position, but another alternative would be to use optimization techniques such as simulated annealing [33].

Coordinate system stitching Some methods focus on fusing the precision of centralized schemes with the computational advantages of distributed systems as we have seen in Sec. 9.3.2.2. Another approach with the same goal that has received some attention [12, 50, 52, 57] is *Coordinate system stitching.* Coordinates system stitching works as follows:

- First, it localizes clusters in the network. They normally are overlapping regions composed by a single node and their one-hop neighbors.
- Then, it refines the localization of the clusters with an optional local map for each cluster placing cluster nodes in a relative coordinate system.
- Finally, it merges those cluster regions computing coordinate transformations between these local coordinate systems.

The fist two steps may be slightly different depending on the algorithm, while the last third step is usually the same. In [50] sub-regions are formed using one-hop neighbors. Then, local maps are computed by choosing three nodes to define a relative coordinate system and using multilateration to iteratively add additional nodes to the map, resulting in a *multilateration sub-tree.*

More robust local maps can be obtained according to [52]. Instead of using three arbitrary nodes to define a map, *robust quadrilaterals* are used, considering a robust quad as a fully-connected set of four nodes where each sub-triangle is also *robust.* A robust sub-triangle with a shortest side of length b and a smallest angle θ must accomplish Eq. (9.50),

$$b \sin^2 \theta > d_{min}, \tag{9.50}$$

where d_{min} is a predetermined constant based on the average measured error. The idea is that the points of a robust quad can be placed correctly with respect to each other. Once an initial robust quad has been chosen, any node that connects to three of the four points in the initial quad can be added using multilateration. This preserves the probabilistic guarantees provided by the initial robust quad, since the node form a new robust quad with the points from the original. By induction, any number of nodes can be added to the local map, as long as each node has a range to three members of the map. These local maps or *clusters*, are now ready to be stitched together.

Coordinates system stitching techniques are quite interesting since they are inherently distributed and they enable the use of sophisticated local maps algorithms. Nevertheless, registering local maps iteratively, can lead to error propagation and perhaps unacceptable error rates as the network grows. In addition, the algorithm may converge slowly since a single coordinate system must propagate from its source to the entire network. Furthermore, these techniques are prone to leave *orphan nodes* because, either they could not be added to the local map, or their local map failed to overlap with neighboring local maps.

9.4 Other Issues in Localization

In this section we outline some aspects involved in the localization theory of wireless ad-hoc and sensor networks that have not been covered in previous sections such as hybrid solutions, mobility and the application of the graph theory.

9.4.1 *Graph theory and localizability*

A fundamental question in the wireless sensor network (WSN) localization is whether a solution to the localization problem is unique. The network, with the given set of anchors, non-anchors and inter-sensor measurements, is said to be *uniquely localizable* if there is a unique set of locations consistent with the given data. Graph theory has been found to be particularly useful for solving the above problem of unique localization. Graph theory also forms the basics of many localization algorithms, especially for the category of distance based localization problem, although it has been used to other types of measurements as well.

A graphical mode for distance based localization problem can be built by representing each sensor in the network uniquely by a vertex. An edge exits between two devices if the distance between the corresponding sensors is known. Note that there is always a vertex between two anchors since the distance can be obtained form their known locations. The obtained graph $G(V, E)$, where V is the set of wireless communication devices and E the set of edges, is called the underlying graph of the sensor network. Details of the graph theoretical representations of the WSN and their use in localization can be found in [28, 43].

9.4.2 *Hybrid schemes*

Hybrid schemes simply combine two or more existing techniques to achieve a better performance such as using both multidimensional scaling (MDS) and proximity based maps (PDS) [13]. Initially, some anchors are deployed (primary anchors). In the first phase some sensors are selected as secondary anchors which are localized thought MDS (Sec. 9.3.1.1). Nodes which are neither primary nor secondary are called normal sensors. In a second phase those normal sensors are localized through proximity distance mapping. Other examples of hybrid schemes are the use of MDS and Adhoc positioning system (APS) [2] and stochastic approaches based on the combination of deductive and inductive methods [44].

9.4.3 *Mobility*

Mobility of sensors nodes obviously have an impact on the localization process. The uncertainty of the node movement may lead to increase the difficulty of the localization task. Nevertheless, in some cases, statistical approaches having capabilities to handle uncertainty of node movements, can tackle localization of mobile sensor nodes. The *sequential Monte Carlo localization (MCL)* method [27] exploits mobility to improve the accuracy and the precision of the localization. The simultaneous localization and tracking scheme based on Laplace method *(LaSLAT)* [81] employs Bayesian filters to accomplish the task of localizing mobile nodes, in which location estimates are iteratively updated given batches of new measurements. Empirical studies have shown that LaSLAT can tolerate noisy range measurements and achieve satisfactory location accuracy.

The localization of static sensors using one mobile anchor equipped with GPS has also been proposed [79]. The mobile anchor periodically transmits a beacon message including its latest position while traversing the area where static sensor nodes are deployed. Upon receiving the beacon packets, a static sensor determines its location relative to the anchor according to the received signal strength (RSS) of the beacon packet through Bayesian inference. The *on beacon mobility scheduling* is also subject of study [38] in order to determine the best beacon trajectory so that each sensor receives sufficient beacon signals with minimum delay.

References

[1] Ahmed, A. A., Li, X., Shang, Y. and Shi, H. (2009). Mds-based localization, *Localization Algorithms and Strategies for Wireless Sensor Networks*, p. 198.

[2] Ahmed, A. A., Shi, H. and Shang, Y. (2005). Sharp: A new approach to relative localization in wireless sensor networks, in *Distributed Computing Systems Workshops, 2005. 25th IEEE International Conference on* (IEEE), pp. 892–898, URL http://ieeexplore.ieee.org/stamp/stamp.jsp?tp=&arnumber=1437278&isnumber=30953.

[3] Akyildiz, I. F., Su, W., Sankarasubramaniam, Y. and Cayirci, E. (2002). Wireless sensor networks: a survey, *Computer networks* **38**, 4, pp. 393–422, URL http://dx.doi.org/10.1016/S1389-1286(01)00302-4.

[4] Bachrach, J. and Taylor, C. (2005). Localization in sensor networks, *Handbook of sensor networks: Algorithms and Architectures* **1**, URL http://people.csail.mit.edu/jrb/Projects/poschap.pdf.

[5] Bahl, P. and Padmanabhan, V. N. (2000). Radar: An in-building rf-based user location and tracking system, in *INFOCOM 2000. Nineteenth Annual Joint Conference of the IEEE Computer and Communications Societies. Proceedings. IEEE*, Vol. 2 (IEEE), pp. 775–784, URL http://ieeexplore.ieee.org/stamp/stamp.jsp?tp=&arnumber=832252&isnumber=18009.

[6] Behnke, R. and Timmermann, D. (2008). Awcl: adaptive weighted centroid localization as an efficient improvement of coarse grained localization, in *Positioning, Navigation and Communication, 2008. WPNC 2008. 5th Workshop on* (IEEE), pp. 243–250, URL http://ieeexplore.ieee.org/stamp/stamp.jsp?tp=&arnumber=4510381&isnumber=4510344.

[7] Blumenthal, J., Grossmann, R., Golatowski, F. and Timmermann, D. (2007). Weighted centroid localization in zigbee-based sensor networks, in *Intelligent Signal Processing, 2007. WISP 2007. IEEE International Symposium on* (IEEE), pp. 1–6, URL http://ieeexplore.ieee.org/stamp/stamp.jsp?tp=&arnumber=4447528&isnumber=4447490.

[8] Blumenthal, J., Reichenbach, F. and Timmermann, D. (2005). Position estimation in ad hoc wireless sensor networks with low complexity, in *Joint 2nd Workshop on Positioning, Navigation and Communication*, pp. 41–49, URL http://www.wpnc.net/fileadmin/WPNC05/Proceedings/Position_Estimation_in_Ad_hoc_Wireless_Sensor_Networks_with_Low_Compexity.pdf.

[9] Bulusu, N., Heidemann, J. and Estrin, D. (2000). Gps-less low-cost outdoor localization for very small devices, *Personal Communications, IEEE* **7**, 5, pp. 28–34, URL http://ieeexplore.ieee.org/stamp/stamp.jsp?tp=&arnumber=878533&isnumber=19016.

[10] Bulusu, N., Heidemann, J. and Estrin, D. (2001). Adaptive beacon placement, in *Distributed Computing Systems, 2001. 21st International Confer-*

ence on. (IEEE), pp. 489–498, URL http://ieeexplore.ieee.org/stamp/stamp.jsp?tp=&arnumber=918979&isnumber=19871.

[11] Bulusu, N., Heidemann, J., Estrin, D. and Tran, T. (2004). Self-configuring localization systems: Design and experimental evaluation, *ACM Transactions on Embedded Computing Systems (TECS)* **3**, 1, pp. 24–60, URL http://dl.acm.org/ft_gateway.cfm?id=972630&type=pdf&CFID=200092476&CFTOKEN=96761448.

[12] Capkun, S., Hamdi, M. and Hubaux, J.-P. (2001). Gps-free positioning in mobile ad-hoc networks, in *System Sciences, 2001. Proceedings of the 34th Annual Hawaii International Conference on* (IEEE), pp. 10–pp, URL http://ieeexplore.ieee.org/stamp/stamp.jsp?tp=&arnumber=927202&isnumber=20032.

[13] Cheng, K.-Y., Lui, K.-S. and Tam, V. (2007). Localization in sensor networks with limited number of anchors and clustered placement, in *Wireless Communications and Networking Conference, 2007. WCNC 2007. IEEE* (IEEE), pp. 4425–4429, URL http://ieeexplore.ieee.org/stamp/stamp.jsp?tp=&arnumber=4225051&isnumber=4224245.

[14] Dil, B. and Havinga, P. (2011). Stochastic radio interferometric positioning in the 2.4 ghz range, in *Proceedings of the 9th ACM Conference on Embedded Networked Sensor Systems (SenSys 11)* (ACM), pp. 108–120, URL http://doi.acm.org/10.1145/2070942.2070954.

[15] Doherty, L., El Ghaoui, L. *et al.* (2001). Convex position estimation in wireless sensor networks, in *INFOCOM 2001. Twentieth Annual Joint Conference of the IEEE Computer and Communications Societies. Proceedings. IEEE*, Vol. 3 (IEEE), pp. 1655–1663, URL http://ieeexplore.ieee.org/stamp/stamp.jsp?tp=&arnumber=916662&isnumber=19795.

[16] Duda, R. O., Hart, P. E. *et al.* (1973). *Pattern classification and scene analysis*, Vol. 3 (Wiley New York).

[17] Farahani, S. (2008). *ZigBee wireless networks and transceivers* (Newnes), URL http://www.learnzigbee.com/.

[18] Fink, A. and Beikirch, H. (2011). Analysis of rss-based location estimation techniques in fading environments, in *2nd IEEE International Conference on Indoor Positioning and Indoor Navigation (IPIN)*, pp. 1–6, URL http://ieeexplore.ieee.org/stamp/stamp.jsp?tp=&arnumber=6071943&isnumber=6071904.

[19] Fink, A., Beikirch, H. and Voss, M. (2010). Improved indoor localization with diversity and filtering based on received signal strength measurements, *International Journal of Computing* **9**, 1, pp. 9–15, URL http://archive.nbuv.gov.ua/portal/natural/computing/2010_1/PDF/10AFRSSM.pdf.

[20] Gezici, S. (2008). A survey on wireless position estimation, *Wireless Personal Communications* **44**, 3, pp. 263–282, URL http://dx.doi.org/10.1007/s11277-007-9375-z.

[21] Gezici, S., Kobayashi, H. and Poor, H. V. (2003). A new approach to mobile position tracking, in *Proc. IEEE Sarnoff Symp. Advances in Wired and Wireless Communications*, pp. 204–207.

[22] Gezici, S. and Poor, H. V. (2009). Position estimation via ultra-wide-band signals, *Proceedings of the IEEE* **97**, 2, pp. 386–403,

URL http://ieeexplore.ieee.org/stamp/stamp.jsp?tp=&arnumber=4796279&isnumber=4802183.

[23] Gezici, S., Tian, Z., Giannakis, G. B., Kobayashi, H., Molisch, A. F., Poor, H. V. and Sahinoglu, Z. (2005). Localization via ultra-wideband radios: a look at positioning aspects for future sensor networks, *Signal Processing Magazine, IEEE* **22**, 4, pp. 70–84, URL http://ieeexplore.ieee.org/stamp/stamp.jsp?tp=&arnumber=1458289&isnumber=31384.

[24] GmbH, N. T. (2007). Real time location systems (rtls), White Paper, NA-06-0148-0391-1.02, URL http://www.nanotron.com/EN/pdf/WP_RTLS.pdf.

[25] Gu, Y., Lo, A. and Niemegeers, I. (2009). A survey of indoor positioning systems for wireless personal networks, *Communications Surveys & Tutorials, IEEE* **11**, 1, pp. 13–32, URL http://ieeexplore.ieee.org/stamp/stamp.jsp?tp=&arnumber=4796924&isnumber=4796921.

[26] He, T., Huang, C., Blum, B. M., Stankovic, J. A. and Abdelzaher, T. (2003). Range-free localization schemes for large scale sensor networks, in *Proceedings of the 9th annual international conference on Mobile computing and networking* (ACM), pp. 81–95, URL http://doi.acm.org/10.1145/938985.938995.

[27] Hu, L. and Evans, D. (2004). Localization for mobile sensor networks, in *Proceedings of the 10th annual international conference on Mobile computing and networking* (ACM), pp. 45–57, URL http://doi.acm.org/10.1145/1023720.1023726.

[28] Jackson, B. and Jordán, T. (2009). Graph theoretic techniques in the analysis of uniquely localizable sensor networks, *Localization algorithms and strategies for wireless sensor networks, G. Mao, B. Fidan (eds), IGI Global,* pp. 146–173URL http://www.maths.qmul.ac.uk/~bill/surveyword.pdf.

[29] Karapistoli, E., Pavlidou, F.-N., Gragopoulos, I. and Tsetsinas, I. (2010). An overview of the ieee 802.15.4a standard, *Communications Magazine, IEEE* **48**, 1, pp. 47–53, URL http://ieeexplore.ieee.org/stamp/stamp.jsp?tp=&arnumber=5394030&isnumber=5394016.

[30] Kay, S. M. (1993). *Fundamentals of Statistical Signal Processing, Volume I: Estimation Theory (v. 1)* (Prentice Hall).

[31] Kim, C., Chin, F. and Garg, H. (2006). Multiple frequencies for accuracy improvement in near field electromagnetic ranging (nfer), in *Personal, Indoor and Mobile Radio Communications, 2006 IEEE 17th International Symposium on* (IEEE), pp. 1–5, URL http://ieeexplore.ieee.org/stamp/stamp.jsp?tp=&arnumber=4022272&isnumber=4022244.

[32] Kim, C. W., Chin, F. and Garg, H. (2007). Selection of frequency for near field electromagnetic ranging (nfer) based on its cramer–rao bound, *Signal Processing Letters, IEEE* **14**, 12, pp. 1000–1003, URL http://ieeexplore.ieee.org/stamp/stamp.jsp?tp=&arnumber=4358007&isnumber=4380451.

[33] Kirkpatrick, S., Jr., D. G. and Vecchi, M. P. (1983). Optimization by simulated annealing, *science* **220**, 4598, pp. 671–680, URL http://www.fisica.uniud.it/~ercolessi/MC/kgv1983.pdf.

[34] Kleinrock, L. and Silvester, J. (1978). Optimum transmission radii for packet radio networks or why six is a magic number, in *Proceedings of*

the IEEE National Telecommunications Conference, Vol. 4 (Birimingham, Alabama), pp. 1–4, URL http://adsabs.harvard.edu/cgi-bin/nph-bib_query?bibcode=1978ntc.....1....4K.

[35] Knapp, C. and Carter, G. (1976). The generalized correlation method for estimation of time delay, *Acoustics, Speech and Signal Processing, IEEE Transactions on* **24**, 4, pp. 320–327, URL http://ieeexplore.ieee.org/stamp/stamp.jsp?tp=&arnumber=1162830&isnumber=26119.

[36] Krishnan, P., Krishnakumar, A., Ju, W.-H., Mallows, C. and Gamt, S. (2004). A system for lease: Location estimation assisted by stationary emitters for indoor rf wireless networks, in *INFOCOM 2004. Twenty-third AnnualJoint Conference of the IEEE Computer and Communications Societies*, Vol. 2 (IEEE), pp. 1001–1011, URL http://ieeexplore.ieee.org/stamp/stamp.jsp?tp=&arnumber=1356987&isnumber=29790.

[37] Kusy, B. and Sallai, J. (2006). Analytical solution for radio-interferometric localization of mobile sensors, Tech. Rep. ISIS-06-710, ISIS, vanderbilt, URL http://www.isis.vanderbilt.edu.

[38] Li, X., Mitton, N., Simplot-Ryl, I. and Simplot-Ryl, D. (2011). Mobile-beacon assisted sensor localization with dynamic beacon mobility scheduling, in *Mobile Adhoc and Sensor Systems (MASS), 2011 IEEE 8th International Conference on* (IEEE), pp. 490–499, URL http://ieeexplore.ieee.org/stamp/stamp.jsp?tp=&arnumber=6076647&isnumber=6076571.

[39] Lin, T.-N. and Lin, P.-C. (2005). Performance comparison of indoor positioning techniques based on location fingerprinting in wireless networks, in *Wireless Networks, Communications and Mobile Computing, 2005 International Conference on*, Vol. 2 (IEEE), pp. 1569–1574, URL http://ieeexplore.ieee.org/stamp/stamp.jsp?tp=&arnumber=1549647&isnumber=33023.

[40] Liu, B.-C., Lin, K.-H. and Wu, J.-C. (2006). Analysis of hyperbolic and circular positioning algorithms using stationary signal-strength-difference measurements in wireless communications, *Vehicular Technology, IEEE Transactions on* **55**, 2, pp. 499–509, URL http://ieeexplore.ieee.org/stamp/stamp.jsp?tp=&arnumber=1608630&isnumber=33784.

[41] Liu, C. and Wu, K. (2005). Performance evaluation of range-free localization methods for wireless sensor networks, in *Performance, Computing, and Communications Conference, 2005. IPCCC 2005. 24th IEEE International* (IEEE), pp. 59–66, URL http://ieeexplore.ieee.org/stamp/stamp.jsp?tp=&arnumber=1460516&isnumber=31417.

[42] Liu, C., Wu, K. and He, T. (2004). Sensor localization with ring overlapping based on comparison of received signal strength indicator, in *Mobile Ad-hoc and Sensor Systems, 2004 IEEE International Conference on* (IEEE), pp. 516–518, URL http://ieeexplore.ieee.org/stamp/stamp.jsp?tp=&arnumber=1392193&isnumber=30305.

[43] Liu, Y., Yang, Z., Wang, X. and Jian, L. (2010). Location, localization, and localizability, *Journal of Computer Science and Technology* **25**, 2, pp. 274–297, URL http://dx.doi.org/10.1007/s11390-010-9324-2.

[44] Lloret, J., Tomas, J., Garcia, M. and Canovas, A. (2009). A hybrid stochastic approach for self-location of wireless sensors in indoor environments, *Sensors* **9**, 5, pp. 3695–3712, URL http://www.mdpi.com/1424-8220/9/5/3695.

[45] Mailaender, L. (2008). On the geolocation bounds for round-trip time-of-arrival and all non-line-of-sight channels, *EURASIP Journal on Advances in Signal Processing* **2008**, p. 37, URL http://dx.doi.org/10.1155/2008/584670.

[46] Mandal, A., Lopes, C. V., Givargis, T., Haghighat, A., Jurdak, R. and Baldi, P. (2005). Beep: 3d indoor positioning using audible sound, in *Consumer Communications and Networking Conference, 2005. CCNC. 2005 Second IEEE* (IEEE), pp. 348–353, URL http://ieeexplore.ieee.org/stamp/stamp.jsp?tp=&arnumber=1405195&isnumber=30469.

[47] Mao, G., Fidan, B. and Anderson, B. (2007). Wireless sensor network localization techniques, *Computer Networks* **51**, 10, pp. 2529–2553, URL http://www.sciencedirect.com/science/article/pii/S1389128606003227.

[48] Maróti, M., Völgyesi, P., Dóra, S., Kusỳ, B., Nádas, A., Lédeczi, Á., Balogh, G. and Molnár, K. (2005). Radio interferometric geolocation, in *Proceedings of the 3rd international conference on Embedded networked sensor systems* (ACM), pp. 1–12, URL http://doi.acm.org/10.1145/1098918.1098920.

[49] McGuire, M., Plataniotis, K. N. and Venetsanopoulos, A. N. (2003). Location of mobile terminals using time measurements and survey points, *Vehicular Technology, IEEE Transactions on* **52**, 4, pp. 999–1011, URL http://ieeexplore.ieee.org/stamp/stamp.jsp?tp=&arnumber=1224556&isnumber=27492.

[50] Meertens, L. and Fitzpatrick, S. (2004). The distributed construction of a global coordinate system in a network of static computational nodes from inter-node distances, *Kestrel Institute TR KES. U* **4**, URL https://www.kestrel.edu/home/people/fitzpatrick/pub/KES.U.04.04.pdf.

[51] Merhi, Z., Elgamel, M. and Bayoumi, M. (2009). A lightweight collaborative fault tolerant target localization system for wireless sensor networks, *Mobile Computing, IEEE Transactions on* **8**, 12, pp. 1690–1704, URL http://ieeexplore.ieee.org/stamp/stamp.jsp?tp=&arnumber=4912210&isnumber=5291226.

[52] Moore, D., Leonard, J., Rus, D. and Teller, S. (2004). Robust distributed network localization with noisy range measurements, in *Proceedings of the 2nd international conference on Embedded networked sensor systems* (ACM), pp. 50–61, URL http://doi.acm.org/10.1145/1031495.1031502.

[53] Nagpal, R. (1999). Organizing a global coordinate system from local information on an amorphous computer, URL http://hdl.handle.net/1721.1/5926.

[54] Navidi, W., Murphy, W. S. and Hereman, W. (1998). Statistical methods in surveying by trilateration, *Computational statistics & data analysis* **27**, 2, pp. 209–227, URL http://dx.doi.org/10.1016/S0167-9473(97)00053-4.

[55] Nerguizian, C., Despins, C. and Affès, S. (2006). Geolocation in mines with an impulse response fingerprinting technique and neural networks, *Wireless Communications, IEEE Transactions on* **5**, 3, pp. 603–611, URL http://ieeexplore.ieee.org/stamp/stamp.jsp?tp=&arnumber=1611090&isnumber=33837.

[56] Niculescu, D. (2004). Positioning in ad hoc sensor networks, *Network, IEEE* **18**, 4, pp. 24–29, URL http://ieeexplore.ieee.org/stamp/stamp.jsp?tp=&arnumber=1316758&isnumber=29182.

[57] Niculescu, D. and Nath, B. (2001). Ad hoc positioning system (aps), in *Global Telecommunications Conference, 2001. GLOBECOM'01. IEEE*, Vol. 5 (IEEE), pp. 2926–2931, URL http://ieeexplore.ieee.org/stamp/stamp.jsp?tp=&arnumber=965964&isnumber=20836.

[58] Pal, A. (2010). Localization algorithms in wireless sensor networks: Current approaches and future challenges, *Network Protocols and Algorithms* **2**, 1, pp. 45–73, URL http://www.macrothink.org/journal/index.php/npa/article/view/279.

[59] Patwari, N., Hero III, A. O., Perkins, M., Correal, N. S. and O'dea, R. J. (2003). Relative location estimation in wireless sensor networks, *Signal Processing, IEEE Transactions on* **51**, 8, pp. 2137–2148, URL http://ieeexplore.ieee.org/stamp/stamp.jsp?tp=&arnumber=1212671&isnumber=27276.

[60] Peng, C., Shen, G., Zhang, Y., Li, Y. and Tan, K. (2007). Beepbeep: a high accuracy acoustic ranging system using cots mobile devices, in *Proceedings of the 5th international conference on Embedded networked sensor systems* (ACM), pp. 1–14, URL http://doi.acm.org/10.1145/1322263.1322265.

[61] Prasithsangaree, P., Krishnamurthy, P. and Chrysanthis, P. (2002). On indoor position location with wireless lans, in *Personal, Indoor and Mobile Radio Communications, 2002. The 13th IEEE International Symposium on*, Vol. 2 (IEEE), pp. 720–724, URL http://ieeexplore.ieee.org/stamp/stamp.jsp?tp=&arnumber=1047316&isnumber=22451.

[62] Priyantha, N. B., Balakrishnan, H., Demaine, E. and Teller, S. (2003). Anchor-free distributed localization in sensor networks, in *Proceedings of the 1st international conference on Embedded networked sensor systems* (ACM), pp. 340–341, URL http://ahvaz.ist.unomaha.edu/azad/temp/ali/03-priyantha-localization-wireless-sensor-networks-cricket.pdf.

[63] Priyantha, N. B., Chakraborty, A. and Balakrishnan, H. (2000). The cricket location-support system, in *Proceedings of the 6th annual international conference on Mobile computing and networking* (ACM), pp. 32–43, URL http://doi.acm.org/10.1145/345910.345917.

[64] Qi, Y. (2003). *Wireless geolocation in a non-line-of-sight environment*, Ph.D. thesis, Princeton University, URL http://cs.gmu.edu/~iyoun/geo/pdf/qi_thesis.pdf.

[65] Rabaey, C. S. J. and Langendoen, K. (2002). Robust positioning algorithms for distributed ad-hoc wireless sensor networks, in *Proceedings of the General Track of the annual conference on USENIX Annual Technical Conference*, pp. 317–327, URL http://portal.acm.org/citation.cfm?id=713854.

[66] Rappaport, T. S. (2001). *Wireless communications: principles and practice*, 2nd edn. (Prentice Hall PTR).

[67] Rappaport, T. S., Reed, J. H. and Woerner, B. D. (1996). Position location using wireless communications on highways of the future, *Communications Magazine, IEEE* **34**, 10, pp. 33–41, URL http://ieeexplore.ieee.org/stamp/stamp.jsp?tp=&arnumber=544321&isnumber=11854.

[68] Römer, K. (2003). The lighthouse location system for smart dust, in *Proceedings of the 1st international conference on Mobile systems, applications and services* (ACM), pp. 15–30, URL http://doi.acm.org/10.1145/1066116.1189036.

[69] Roos, T., Myllymaki, P. and Tirri, H. (2002). A statistical modeling approach to location estimation, *Mobile Computing, IEEE Transactions on* **1**, 1, pp. 59–69, URL http://ieeexplore.ieee.org/stamp/stamp.jsp?tp=&arnumber=1011059&isnumber=21786.

[70] Sahinoglu, Z. and Gezici, S. (2006). Ranging in the ieee 802.15.4a standard, in *Wireless and Microwave Technology Conference, 2006. WAMICON'06. IEEE Annual* (IEEE), pp. 1–5, URL http://ieeexplore.ieee.org/stamp/stamp.jsp?tp=&arnumber=4161059&isnumber=4141009.

[71] Sandor Szilvasi, P. V. (2012). *Interferometry - Research and Applications in Science and Technology*, chap. Interferometry in Wireless Sensor Networks (InTech), pp. 32–43, URL http://www.intechopen.com/books/interferometry-research-and-applications-in-science-and-technology/interferometry-in-wireless-sensor-networks.

[72] Savarese, C., Rabaey, J. M. and Beutel, J. (2001). Location in distributed adhoc wireless sensor networks, in *Acoustics, Speech, and Signal Processing, 2001. Proceedings.(ICASSP'01). 2001 IEEE International Conference on*, Vol. 4 (IEEE), pp. 2037–2040, URL http://ieeexplore.ieee.org/stamp/stamp.jsp?tp=&arnumber=940391&isnumber=20357.

[73] Savvides, A., Srivastava, M., Girod, L. and Estrin, D. (2004). Localization in sensor networks, *Wireless sensor networks*, pp. 327–349URL http://dl.acm.org/citation.cfm?id=1013825.1013841.

[74] Schantz, H. and Fluhler, J. (2006). Near-field technology–an emerging rf discipline, in *The European Conference on Antennas and Propagation: EuCAP 2006*, Vol. 626, p. 2, URL http://www.q-track.com/documentation/IEEE-APS-2006.pdf.

[75] Schantz, H. G. (2005). Near field phase behavior, in *Antennas and Propagation Society International Symposium, 2005 IEEE*, Vol. 3 (IEEE), pp. 134–137, URL http://ieeexplore.ieee.org/stamp/stamp.jsp?tp=&arnumber=1552452&isnumber=33059.

[76] Schantz, H. G. (2007). A real-time location system using near-field electromagnetic ranging, in *Antennas and Propagation Society International Symposium, 2007 IEEE* (IEEE), pp. 3792–3795, URL http://ieeexplore.ieee.org/stamp/stamp.jsp?tp=&arnumber=4396365&isnumber=4395410.

[77] Shang, Y., Rumi, W., Zhang, Y. and Fromherz, M. (2004). Localization from connectivity in sensor networks, *Parallel and Distributed Systems, IEEE Transactions on* **15**, 11, pp. 961–974, URL http://ieeexplore.ieee.org/stamp/stamp.jsp?tp=&arnumber=1339247&isnumber=29526.

[78] Shang, Y., Ruml, W., Zhang, Y. and Fromherz, M. P. (2003). Localization from mere connectivity, in *Proceedings of the 4th ACM international symposium on Mobile ad hoc networking & computing* (ACM), pp. 201–212, URL http://doi.acm.org/10.1145/778415.778439.

[79] Sichitiu, M. L. and Ramadurai, V. (2004). Localization of wireless sensor networks with a mobile beacon, in *Mobile Ad-hoc and*

Sensor Systems, 2004 IEEE International Conference on (IEEE), pp. 174–183, URL http://ieeexplore.ieee.org/stamp/stamp.jsp?tp=&arnumber=1392104&isnumber=30305.

[80] Tarrío, P., Bernardos, A. M. and Casar, J. R. (2011). Weighted least squares techniques for improved received signal strength based localization, *Sensors* **11**, 9, pp. 8569–8592, URL http://www.mdpi.com/1424-8220/11/9/8569/pdf.

[81] Taylor, C., Rahimi, A., Bachrach, J., Shrobe, H. and Grue, A. (2006). Simultaneous localization, calibration, and tracking in an ad hoc sensor network, in *Proceedings of the 5th international conference on Information processing in sensor networks* (ACM), pp. 27–33, URL http://doi.acm.org/10.1145/1127777.1127785.

[82] VanTrees, H. L. (1968). *Detection, Estimation, and Modulation Theory(Part I)* (John Wiley and Sons).

[83] Wang, J., Ghosh, R. and Das, S. K. (2010). A survey on sensor localization, *Journal of Control Theory and Applications* **8**, 1, pp. 2–11, URL http://dx.doi.org/10.1007/s11768-010-9187-7.

[84] Wang, X., Moran, B. and Brazil, M. (2007). Hyperbolic positioning using rips measurements for wireless sensor networks, in *Networks, 2007. ICON 2007. 15th IEEE International Conference on* (IEEE), pp. 425–430, URL http://ieeexplore.ieee.org/stamp/stamp.jsp?tp=&arnumber=4444124&isnumber=4444031.

Chapter 10

Mobile Wireless Sensor Networks: Challenges and Business Applications

Essia Hamouda[1] and John Gerdes[2]
[1] *California State University Fullerton, USA,*
[2] *University of South Carolina, Integrated Information Technology, USA*

Abstract. *Wireless sensor networks (WSNs) have emerged as an effective solution for a wide range of business applications. Most of the traditional WSN architectures consist of stationary nodes that are densely deployed over a sensing area. Recently, several WSN architectures deploying mobile devices have been proposed to exploit node mobility, and address problems of data transmission and collection in WSNs. In this chapter, we first define mobile wireless sensor networks (MWSNs). Then, we present an overview of the major challenges encountered in data transmission in MWSNs. Finally, we focus on the importance of MWSNs in business applications. We identify recent business applications where mobile wireless sensors have become important along with the challenges they are facing.*

10.1 Adding Mobility to Wireless Sensor Networks

In recent years, Wireless Sensor Networks (WSNs) have become an established technology for a large number of applications. A sampling of these applications include:

- environmental monitoring (e.g., pollution prevention, structure and building health)
- event detection (e.g., intrusion, fire/flood emergencies)
- target tracking (e.g., surveillance)
- agriculture (e.g., precision agriculture)

- travel (e.g., congestion monitoring)
- retail (e.g., inventory control and monitoring product quality)
- industry (e.g., inventory movement, process monitoring, maintenance).

WSNs are networks of autonomous, energy-constrained nodes that perform three basic tasks: (i) collect data from the surrounding environment, (ii) process (and possibly store) the acquired data, and (iii) transfer them through wireless communications to a data collection point called sink node or base station.

Figure 10.1 shows a typical Mobile Wireless Sensor Network (MWSN) configuration. Note that the network density can vary, with the top portion of the network having a dense deployment of sensor resulting in good connectivity while the bottom portion is sparsely populated which would likely result in portions of the network being disconnected and unable to maintain a connection to the data sink. Both mobile and stationary sensor nodes and data sinks are depicted in the network. The bottom event detection occurs in the sparsely populated portion of the network. If a connection cannot be obtained immediately, the sensor node would need to locally store the collected process data and wait until a connection is possible. However, if the data is urgent (case of a fire for example) the wait should not be long. In the case illustrated, a mobile data sink has moved within communication range enabling the node to gain a connection and report the detected event.

WSNs are related to radio frequency identification (RFID) tags that have been used for years to identify objects within a network. RFID tags are typically not considered to be part of a sensor network. While there are similarities, there are also key differences. Both sensors and RFID tags do have on-board storage, but RFID tags store static information used to identify the tag's attached entity in contrast to the data collected by the sensors from their local surroundings. Unlike WSN sensors, RFID tags do not incorporate sensors that monitor the local environment, and do not have on-board processing capability.

The traditional WSN architectures are based on the assumption that the network is dense, so that any two devices communicate with each other through multi-hop paths. Both the sensors and sinks are assumed to be stationary, and mobility is not considered. Recently researchers have considered the benefits of extending WSNs to incorporate mobile devices. Adding mobility opens up new applications of WSNs. For example, mobile sensors can be used to monitor free-ranging animals, inventory movement in indus-

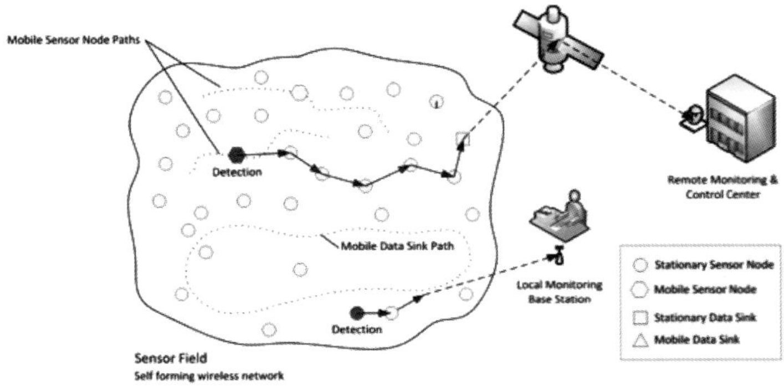

Fig. 10.1 Data transmission in a wireless sensor network.

trial environments, as well as monitoring patient health.

WSN Mobility can be achieved in different ways. The sensing nodes may be stationary, with the data sink moving throughout the network (See Chapter 2. For example, in agricultural settings, a sparse sensor network could be monitoring local environmental conditions, with the data sink mounted on mobile irrigation systems, or farm vehicles that routinely move throughout the network providing intermittent connection to the stationary remote sensors. Alternatively, sensors may be mounted on free-roaming livestock that move throughout a monitored region equipped with stationary communication sensors that communicate with the sink. When a monitored animal moves within communication range, sensor data can be offloaded and transmitted to the data sink. Sensing nodes may move randomly, or in a prescribed pattern on a predictable schedule. If there is a known movement pattern, this can be used to conserve battery life, with devices turning on only when they can communicate with the mobile nodes. It is also possible for node mobility to be controlled or managed to improve power utilization. Controlled mobile nodes are nodes that are static most of the time and are ordered to move from a given location to another specific one. Once they get to the new location they become static until they receive a future order to move.

10.1.1 *Benefits of mobility in WSN*

Mobility provides a number of key benefits to WSNs. We briefly discuss some of these benefits.

Improved connectivity. In traditional WSNs the network is typically dense to ensure that the network is fully connected. Sensor information is lost in portions of the network that cannot reach a sink node. When node mobility is introduced, network connectivity changes over time resulting in a dynamically connected network. In this way, mobile elements can service isolated regions. In applications where such intermittent connectivity is acceptable, a sparse WSN architecture becomes a feasible option.

Reduced cost. Moving to a less dense mobile WSN means fewer nodes are deployed, subsequently reducing the network cost. Although adding mobility features to the sensors might be expensive, in many cases it is possible to exploit mobile elements already present in the sensing area (e.g., vehicles, moving equipment, mobile personnel [18, 30] and attach sensors or data sinks to these elements.

Increased reliability. In all WSNs the communication paradigm is often ad hoc, involving multi-hops between sensors and data sink. Environmental factors can negatively impact communication reliability. Nearby devices can cause interference, either from radio frequency (RF) noise or from competing communication traffic. Any changes in environmental factors can impact connectivity. Signal degradation could be weather related, or due to signal path blockage resulting from changes in the surrounding layout, or multipath interference. Message loss also increases with the number of hops. Adding mobility may mitigate some of these factors. As nodes move, communication characteristics also change. Alternate pathways may be established around previous barriers. If node movement is predictable, sensors can optimize their energy usage by scheduling their communication with these periodic visits. For example, sensors monitoring remote, inaccessible locations might communicate through an orbiting satellite that passes overhead on a regular basis.

Energy efficiency. The traffic pattern inherent to WSNs is convergecast, meaning that messages are generated from many sensor nodes are routed toward and converge on the data sink. As a consequence, nodes closer to the sink are more active than others, and subject to accelerated energy depletion. This phenomenon is also known as the funneling effect. Sensor device mobility helps reduce this funneling effect, as their movement can cause traffic to be routed through different regions of the network and spread the energy consumption more uniformly, even in the case of a dense WSN architecture [11, 35].

10.1.2 *Issues associated with MWSNs*

Independent of how mobility is achieved, it introduces a number of important issues that impact the performance of the network, factors which do not arise in static WSNs, namely:

Contact detection. Communication is possible only when nodes are within transmission range of each other. As a result it is necessary to efficiently detect the presence of a mobile node. This is especially true when contact duration is short.

Mobility-aware power management. It may be possible to exploit the knowledge of the mobility pattern to further optimize the detection of mobile elements. In fact, if visiting times are known or can be accurately predicted, sensor nodes can conserve energy and awake only when they expect the mobile element to be within their transmission range.

Reliable data transfer. Because mobility might limit the frequency and duration that sensors can connect to the sink, data transfer protocols must be efficient. In addition, since nodes move during data transfer, message exchange protocols must be mobility-aware. In the case of transmission problem, sensor on-board data storage has to be sufficient to capture event data and cache it until it can transmit it to the sink. (See also Chapter 1).

Managed node mobility. When the motion of mobile elements can be controlled, a policy for visiting nodes in the network could be defined to improve performance. To this end, the path and the speed or sojourn time of mobile nodes have to be defined in order to improve (optimize) the network performance.

Loss of, or intermittent network connectivity. Node movement can lead to a disconnected network necessitating that routing protocols must be adaptive and connectivity-aware. Network connectivity issues can also be addressed by deploying dense homogeneous networks to increase resources per unit area. However, dense networks experience other problems, such as device management overhead, increased transmission interference and contention. Another approach is to introduce specialized nodes with long-range communication capabilities into the network to improve network connectivity. Such nodes are, on the other hand, expensive and consume a lot of energy.

Reduced network lifetime. Network lifetime is directly related to how long the power resources in sensor nodes will last. Designing and using energy-efficient protocols and algorithms can increase the network lifetime. One scheduling scheme organizes sensors to work in batches to extend the network life [10]. Another solution is to take advantage of node mobility by relocating them to areas where resources are most needed to efficiently transmit data. It has been shown [36] that deploying resource-rich mobile devices in a network can provide the same performance as increasing the network density.

Communication Interference. A key challenge faced by WSN for some applications is to provide long-term reliable operation under signicant interference and environmental dynamics experienced in real-life environments. Wireless links are vulnerable to interference from various sources. This problem has become more serious with the increased use of the unlicensed 2.4 GHz band. WSN competes with a wide range of wireless devices, such as IEEE 802.11b/g wireless router, Bluetooth headsets, 2.4 GHz cordless phones etc. Compared with these interferers, WSN is in weak position due to its low transmission power and limited hardware resources.

In some installations there can also be interference between different suppliers of WSN systems. In industrial environments, equipment manufacturers often offer process-monitoring systems that are WSN based. These systems monitor both the manufacturing process as well as the operational health of the equipment. Operating WSNs from different vendors in close proximity to one another can cause communication conflicts.

Security. There are severe constraints to incorporating security into a WSN due to limitations in sensor storage, communication, computation, and processing capabilities [32]. Most of the traditional wired network security techniques are unsuitable in low power devices such as wireless sensor networks. This is due largely to the fact that typical key exchange techniques use asymmetric cryptography that is computationally intensive for the individual nodes in a sensor network. Symmetric cryptography is therefore the typical choice for applications that are resource limited.

The traditional example of symmetric cryptography is DES (Data Encryption Standard). The use of DES, however, is quite limited due to the fact that it can be broken relatively easily. In light of the shortcomings of DES, other symmetric cryptography systems have been proposed including 3DES (Triple DES), RC5, AES.

The security of wireless networks is protocol dependent. ANSI/ISA-100.11a and IEEE 802.15.4 protocols use encryption technologies that provide equivalent or higher levels of security than that provided by a wired communication system, protecting data confidentiality, authenticity, and freshness [6]. Zigbee security is based on symmetric keys.

The rest of this chapter is organized as follows. Section 10.2 presents real life business applications where sensor mobile devices are playing a very important role and the challenges facing such applications. Finally, Section 10.3 offers concluding remarks, with directions on open research issues.

10.2 Business Applications of Mobile Wireless Sensor Networks

Today, enterprises in retail, healthcare, finance, government, transportation, industry, agriculture and others are finding new and better paths to maximize customers satisfaction and competitive opportunities by relying on a business infrastructure with robust wireless mobility capability. Generally the business advantages of converged wired, wireless with mobile wireless infrastructures are numerous ranging from flexibility to cost reduction. In the following section we discuss the use of MWSNs within three important business domains (industry/manufacturing, healthcare, and agriculture), along with the challenges associated with those domains, and some illustrative applications where mobile wireless sensor networks have been implemented.

10.2.1 *MWSN industrial applications*

Wired and wireless sensor networks are used throughout Industry for various applications including building automation, process automation, automatic meter reading, inventory management, and intrusion detection. Wired sensor networks are commonly used for process monitoring and control because they provide the necessary security, and communication speed. The flexibility afforded by WSNs is increasing especially applied in time critical applications [7].

The use of wireless sensor networks provides for increased flexibility, and significant cost advantages for the installation and operation of industrial systems [26, 37]. Installation of wireless sensors is up to 10 times cheaper than the wired alternative [23] and simplifies both installation and maintenance [37, 38]. Examples of industrial applications that use WSNs include:

Monitoring environmental conditions. Many companies are subject to regulations related to process emissions. WSN can be quickly deployed to monitor aspects of the environment such as stack emission, and process waste water, and are especially useful for documenting compliance issues in remote and/or hazardous locations. These systems are also useful for security and intrusion detection applications [2].

Condition monitoring. Wireless sensors are used to monitor the operational health of motors, bearings and machine equipment, monitoring for vibration, as well as changes in temperature and pressure, and then transmitting alerts when maintenance is required. [13, 16, 17].

Process automation. Wireless sensors are used to monitor resources used in manufacturing, including raw material, current stock, and supply chain status, as well as the location of key personnel or equipment. WSN can also be useful in process performance monitoring, evaluation and improvement [7]. Wireless sensors can capture various process data including temperature, pressure, humidity, flow, level, viscosity, density and vibration intensity. This information is then transmitted to a control system for operation and management. For example, in chemical plants WSNs are used to monitor chemical storage tanks and other plan operations. If sensors identify an issue that needs to be addressed, alerts are sent through the WSN which triggers messages sent directly to field operators that then have the information to quickly identify the right transmitter and task.

Supporting enterprise mobility is a priority in industry today [3]. The wireless nature of these sensors allows their use in mobile environments [21]. While RFID (radio frequency identification) tags can be attached to products and materials to encode identification information, RFID tags are dumb devices and cannot support the sensing and local analysis features available with wireless sensors. Applications that take advantage of this feature include mobile asset and material tracking. Sensors attached to raw material, work in progress, and current stock can monitor location and status as these parts move through the supply chain. Other sensors can monitor the status and location of equipment operators and management personnel. For example, when dealing with fruit or pharmaceutical products it is important to have a temperature controlled supply chain (cold chain management) including product storage and transport to avoid degradation and spoilage [4, 31]. Wireless sensors can be used to monitor critical environment factors and alert if they are not adequately controlled.

Adding mobile nodes to a WSN increases flexibility and provides process information that would otherwise be difficult to obtain. For example, some products impose strict environmental constraints on their manufacturing, transport and storage. For example, frozen foods must be kept frozen as they move through the supply chain. Some pharmaceuticals are also very temperature sensitive, and require special handling. Mobile sensors can be mounted on these products or shipping containers so they can monitor critical environmental values and send an alert if they deviate from prescribed values. Messages from wireless sensors packed with the products would then link-up through ad-hoc networks to different data sinks or base stations as they move through the supply chain from manufacturer to wholesaler, and finally retailer.

10.2.2 *WSN medical applications*

WSNs have recently made significant inroads with regards to medical monitoring applications including pre-hospital and in-hospital emergency care, home-based monitoring, disaster response, and stroke patient rehabilitation. Numerous applications such as physiology monitoring (i.e., heart rate, blood pressure, and blood oxygen level), using either implantable sensors, endoscopic capsule, or wearable devices are already in use.

They provide a cost effective way to provide remote monitoring of the health of patients. Geisinger Health Plan's home tele-monitoring program helped reduce 30-day hospital readmission rates by 44% among patients with congestive heart failure [9]. Sensors monitoring vital signs can be worn by the individual. Many applications include one or more types of sensors carried by the patient forming what has been referred to as either a body area network (BAN), or a body sensor network (BSN) as shown in Figure 10.2. These sensors monitor the patient's physiological data and vital signs providing an integrated view of the patient's condition. Additionally, one or more types of sensors are deployed within the environment forming a personal area network (PAN). These two networks are wirelessly connected to a backbone network via a gateway node [1] see Figure 10.3. Once received by the host application, patient data is logged, analyzed, and alerts are generated.

This model provides a number of key benefits. It is unobtrusive, only requiring the patient to carry or wear the designated sensors. Monitoring is done based on normal activity. It is automated, and thus the user does not have to remember to record readings, and eliminates transcription errors. Data can be recorded more frequently, providing a clearer picture of the patient's condition. It provides continuous monitoring and can trigger alerts if critical events occur. WSNs provide both cost savings and convenience since the patient does not need to go to a hospital

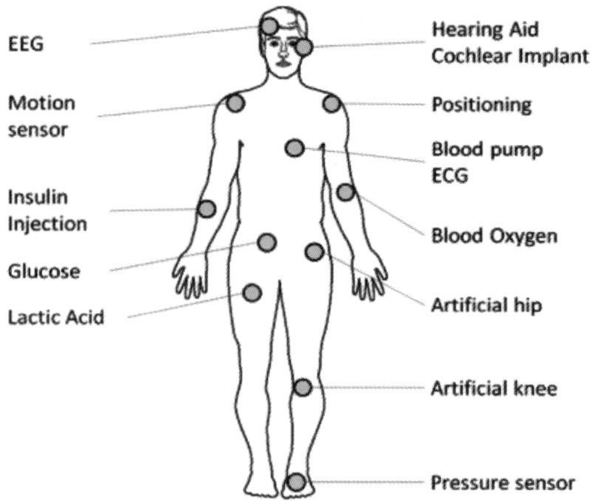

Fig. 10.2 Example of patient monitoring in a Wireless Body Area Network.

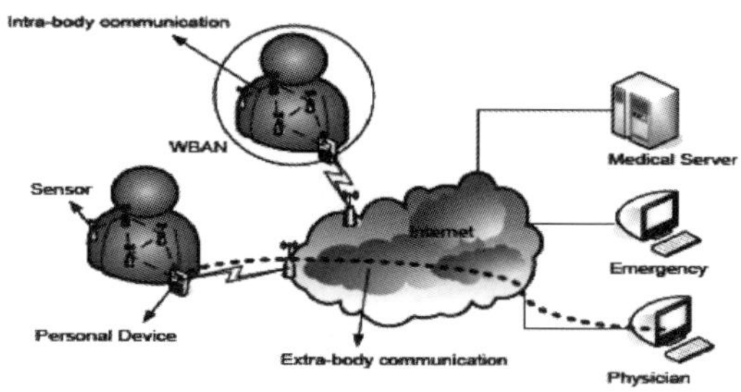

Fig. 10.3 Example of intra-body and extra-body communication in a WBAN.

or doctor's office to have this data collected. Data is automatically captured and transmitted through the WSN to the gateway node, which then transmits this information to the monitoring system through traditional telecommunication networks. Naturally, since these applications deal with health issues, the reliability of both the devices and network are important. Since these devices are dealing with personal health data, privacy and security are important considerations in medical applications [22]. While the need for strong security is universally acknowledged due to the personal and highly sensitive data transmitted, many systems do not adequately address this issue. Sensor characteristics that relate to mobility and portability are critical for medical applications. Since sensors may be carried or worn by patients, so they must be small and lightweight. Low-power radios using the IEEE 802.15.4 standard have been adopted by the most of the available body sensor networks [28]. Power management is critical for these applications as it impacts the long-term monitoring of patients. These competing requirements result in considerable design challenges for sensors used for medical applications. Wireless medical sensors may be wearable, implantable or portable, and integrated on various kinds of wireless communication motes [19]. Tradeoffs are necessary based on the amount and frequency of data to be collected, power utilization, and the need to keep nodes small and lightweight. One approach is to locally process and analyze the data, and then transmit just the critical information rather than the raw data. This approach is constrained by the sensor's on-board processing power. Wireless Sensor Networks have been applied to a wide variety of medical applications. A few of these applications are outlined below. HealthGear uses a blood oximeter to monitor the patient's blood oxygen level and pulse while sleeping for detecting sleep apnea [25]. The sensor connected through Bluetooth to the user's cell phone that stores, analyzes and transmits the data. A set of physiological sensors connected via Bluetooth to a cell phone. It is basically a wearable real-time health system for monitoring and analyzing physiological signals. WSNs have been used to monitor motor fluctuations in patients with Parkinson's Disease using wearable sensors [28]. Accelerometers are used to assess the severity of tremors, and dyskinesia (difficulty in performing voluntary motor movements). By attaching sensor to patients in hospitals, their health status and location can be monitored. If there is a medical emergency, medical staff can be alerted to the problem as well as the location of the patient. Doctors can also carry sensors so that they can be located at all times. Sensors worn by patients can capture vital signs in real-time and transmit this data to handheld computers carried by medical personnel. Sensors can also store patient data such as identification, history, and treatments. Harvard University is working with the School of Medicine at Boston University to develop CodeBlue, a system designed to support wireless medical sensors, PDAs, PCs, and other devices used in the monitoring and treat patients [8]. Pilot studies have focused on addressing the needs of data management in disaster and emergency medical situations where traditional methods can be quickly overwhelmed. One such study developed an extensible medical information tag, body sensor network system which automatically tracks patient medical data throughout the disaster recovery process from the disaster scene, to ambulances, and finally to the hospital [12]. This extensible platform supports various sensor add-ons including GPS, pulse oximetry, blood pressure, temperature, ECG, etc., and relays this data over a self-organizing wireless mesh networks. WSNs can be used to monitor drug administration. Sensor notes attached to medications as well as patients can be used to cross check drug distribution. Sensors can also be used to identify patient allergies and required medications, and minimize drug side effects [24].

10.2.3 *WSN agriculture applications*

The ability of WSN to establish a distributed network to monitor and report on local environmental factors can be utilized in agricultural applications. The combination of sensors to monitor soil moisture, humidity, soil analysis, etc., enables managed agriculture. WSN provides the farmer with detailed profiles that can lead to precision agriculture, where the local environmental needs to dictate the appropriate, targeted application of fertilizer, insecticide and herbicide, thereby eliminating waste and improving yield. It has the side benefit of reducing ground water pollution problems associated with agricultural runoff. Sensors can also trigger an automated call to workers to alert them that the crop is ready to harvest [5].

There are a number of key characteristics that make WSN attractive in agricultural applications. The wireless nature of the sensors is especially beneficial. Agricultural applications are typically outside and can cover relatively large geographic areas, and thus the flexibility these wireless sensors provide is beneficial. The wireless nature eliminates the cabling costs, speeds deployment and support sensor mobility. They have the further benefit that they often can use solar panels to help recharge the on-board battery thereby reducing power management issues. Compared to some other applications, the sensitivity of the data is typically lower, which reduces the importance of security. Should a sensor fail, the network can reconfigure itself to maintain connectivity. Since process times are relatively long, transmission rate is not as critical as some other applications.

Environmental monitoring can benefit from the automated sensing capability of WSNs [36]. As the cost of deployment decreases, it is possible to capture greater detail that improves operations. By geo-coding the data, the region can be monitored and remotely managed to address local needs. This extends to weather, frost prevention, and remotely monitoring crop maturity to determine when it is time to harvest.

Camalie Vineyards installed a network of wireless soil moisture monitors in a portion of their vineyards (4.4 acres), which is managed as 10 distinct zones due to differences in soil makeup, exposure and climate [33]. Water management is critical to the success of managing any agricultural operation. The added visibility provided by the WSN caused changes in operation [14]. The wireless soil moisture monitors allowed Camalie Vineyards to manage their irrigation to keep optimal soil moisture to improve both yield and improved quality. The vineyard has expanded beyond just measuring soil moisture. Using different type of sensors, they track temperature, humidity and water flowing to each vine, and in their production process they measure the wine cellar temperature alone with the rate of fermentation.

The testing of this system was during the 2005–2007 growing seasons. Camalie Vineyards observed good results, with increasing yields and excellent grape quality. The management indicates that the increased visibility of the soil moisture data played a role in these results. Based on the success of the initial test, the Camalie Vineyard has expended soil moisture monitoring to 25 sites covering all of Camile's Vineyards as well as two neighbor vineyards. [14].

WSN have also been used to visually monitor crops for evident of pests or disease [20]. Image processing techniques are used to examine images of the leaves and identify abnormalities.

10.2.4 *Challenges of WSNs operating in a business environment*

Although there are numerous benefits for implementing wireless sensor networks in any business application, there are inherent challenges with wireless systems, which make them less reliable than wired arrangements. Moreover, due to the diversity nature of business applications that utilize WSNs, each wireless sensor network design must contend with different operating constraints. No individual sensor design, or communication protocol is appropriate for all applications.

Three design issues/challenges are often critical in business environments, namely: quality of service (QoS), communication interference, and security. We addressed the latter challenges in Section 10.1 and we discuss QoS below.

Quality of Service Business applications can have dramatically different quality of service requirements. In an industrial application for example, the response time requirements can range from 1 msec for applications related to process control, up to tens of seconds or even longer for those involving supply chain actions [27, 29, 34]. For instance, for a safety-critical control system, large delay in transmitting data from sensors to a sink and packet loss occurring during the course of transmission may not be allowed, while they may be acceptable for an air-conditioning system that maintains the temperature inside an office.

In the presence of resource constraints, the network QoS may suffer from the unavailability of computing and/or communication resources. As an example, a number of nodes that want to transmit messages over the same WSAN have to compete for the limited bandwidth that the network is able to provide. As a consequence, some data transmissions will possibly experience large delays, resulting in low level of QoS. Due to the limited memory size, data packets may be dropped before the nodes successfully send them to the destination. Therefore, it is of critical importance to use the available resources in WSNs in a very efficient way.

Two international standards that deal with WSNs have been developed and approved. WirelessHART was adopted in 2010 as IEC 62591, and a competing standard, ANSI/ISA-100.11a, was adopted in 2011. Both standards strive to provide reliable, secure communication, and are designed to work with energy-constrained devices. According to ISA,

> *the [ISA-100.11a] standard is intended to provide reliable and secure wireless operation for non-critical monitoring, alerting, supervisory control, open loop control, and closed loop control applications. It defines the protocol suite, system management, gateway, and security specifications for low-data-rate wireless connectivity with fixed, portable, and moving devices supporting very limited power consumption requirements. The application focus is to address the performance needs of applications such as monitoring and process control where latencies on the order of* 100 *msec can be tolerated, with optional behavior for shorter latency. [15]*

10.3 Conclusions and Directions for Future Research

In this work, we introduce the reader to mobile wireless sensor networks. Stationary sensor networks have been applied to a broad array of application, and have proven to be a cost effective way to provide system monitoring and control. Recently mobile sensor networks have been proposed to extend the capabilities of WSNs. For example, a wireless mobile sensor can be instructed to move from area to area to be closer to a target or interest, or to adjust for a hole in a disconnected network. Recently, MWSN architectures have been proposed in healthcare, environment monitoring, industry, etc. We have discussed several challenges associated with adding mobility to these wireless networks.

In future work, we plan to focus on designing routing policy that take advantage of sensor devices mobility while ensuring (1) data transmission security to protect the sensitive data relayed by sensor devices, and (2) energy efficiency to guarantee network connectivity, reduce high delays in transmitting urgent data and reduce data loss. In order to be useful in real world applications the routing policy needs to be general enough to allow several sensor nodes to communicate with multiple sinks while addressing the significant challenges of interference and limited power.

References

[1] Alemdar, H. and Ersoy, C. (2010). Wireless sensor networks for healthcare: a survey, *Computer Networks* **54**, p. 2688 – 2710.

[2] Basha, E. A., Ravela, S. and Rus, D. (2008). Model-based monitoring for early warning food detection, in *Proc. of the 6th ACM conference on Embedded network sensor systems (SenSys)* (New York, NY, USA), pp. 295–308.

[3] Beilstein, J. (2012). Owens corning on mobility in manufacturing vision, in *ARC, 2012 World Industry Forum* (Orlando FL., USA), http://www.youtube.com/watchv=H5rtVegpbYE.

[4] Bijwaard, D., Kleunen, W., Havinga, P., Kleiboer, L. and Bijl, M. (2011). Industry: using dynamic wsns in smart logistics for fruits and pharmacy, in *Proc. of the 9th ACM conference on Embedded network sensor systems (SenSys)* (Seattle, WA, USA), pp. 218–231.

[5] Burrell, J., Brooke, T. and Beckwith, R. (2004). Vineyard computing: Sensor networks in agriculture production, in *Proc. of the 2nd IEEE conference on pervasive computing and communications (PerCom)* (Orlando, FL, USA), pp. 38–45.

[6] Cisco and Emerson (2009). Integrating an industrial wireless sensor network with your plants switched ethernet and IP network, .

[7] Erdelj, M., Mitton, N. and Natalizio, E. (2013). Applications of industrial wireless sensor networks, in *Industrial Wireless Sensor Networks: Applications, Protocols, and Standards* (CRC Press), pp. 1–22.

[8] Fulford-Jones, T., Malan, D., Welsh, M. and Moulton, S. (2004). Codeblue: An ad hoc sensor network infrastructure for emergency medical care, in *International Workshop on Wearable and Implantable Body Sensor Networks* (London, UK).

[9] Gale, L. (2013). Telemonitoring lowers readmission rates at GEISINGER, clinical innovation + technology.

[10] Gallais, A., Carle, J., Simplot-Ryl, D., Parvery, H. and Gorce, J.-M. (2006). Efficiency impairment of sensor area coverage protocols under realistic physical layer conditions, in *Proc. of the 10th IEEE Singapore International Con-*

ference on Communication Systems (Singapore), pp. 1–5.

[11] Gandham, S. R., Dawande, M., Pakash, R. and Venkatesan, S. (2003). Energy efficient schemes for wireless sensor networks with multiple mobile base stations, in *Proc. of the Global Telecommunications Conference*, pp. 377–381.

[12] Gao, T., Pesto, C., Selavo, L., Chen, Y., Ko, J. G., Lim, J. H., Terzis, A., Watt, A., Jeng, J., Chen, B., Lorincz, K., and Welsh, M. (2008). Wireless medical sensor networks in emergency response: Implementation and pilot results, in *Proc. of the 8th IEEE International Conference on Technologies for Homeland Security (HST)* (Waltham, MA, USA), pp. 377–381.

[13] Gbur, G., Wier, W., Bark, T., and Baldwin, B. (2006). Wireless vibration monitoring in a us coal-fired plant. *Insight* **48**, 8, p. 456.

[14] Holler, M. (2008). Update Camalie Networks wireless sensing, http://camalie.com/wirelesssensing/wirelesssensors.htm.

[15] ISA (2011). Ansi/isa-100.11a-2011 wireless systems for industrial automation: Process control and related applications, .

[16] Kevan, T. (2006). Shipboard machine monitoring for predictive maintenance, *Wireless Sensors Magazine, Qestex* .

[17] Krishnamurthy, L., Adler, R., Buonadonna, P., Chhabra, J., Flanigan, M., Kushalnagar, N., Nachman, L., and Yarvis, M. (2005). Design and deployment of industrial sensor networks: Experiences from a semiconductor plant and the north sea, in *Proc. of the 3rd ACM conference on Embedded network sensor systems (SenSys)* (San Diego, CA, USA).

[18] Kroc, S. (2003). Personal wireless sensor network for mobile health care monitoring, in *Proc. of the 6th International Conference on Telecommunications in Modern Satellite, Cable and Broadcasting Service (TELSIKS)* (Montenegro), pp. 471–474.

[19] Kumar, P. and Lee, H. J. (2013). Security issues in healthcare applications using wireless medical sensor networks: a survey, *Sensors* **2**, 1, p. 55 – 91.

[20] Lloret, J., Bosch, I., Sendra, S. and Serrano, A. (2011). Wireless sensor network for vineyard monitoring that uses image processing, *Sensors* **11**, 6, p. 6155 – 6196.

[21] Low, K., N, W. W. N. and Meng, J. (2005). Wireless sensor networks for industrial environments, in *Proc. of the International Conference on Computational Modelling, Control and Automation, and International Conference on Intelligent Agents* (Vienna, Austrich), pp. 271–276.

[22] M., A. A., Liu, J. and Kwak, K. (2012). Security and privacy issues in wireless sensor networks for healthcare applications, *Journal of medical systems* **36**, pp. 93–101.

[23] McAdams, B. E. (2012). Wireless technologies for industrial automation, control engineering, http://www.controleng.com/single-article/wireless-technologies-for-industrial-automation/845d0b546e.html.

[24] Noury, N., Herve, T., Rialle, V., Virone, G., Mercier, E., Morey, G., Moro, A., and Porcheron, T. (2000). Monitoring behavior in home using a smart fall sensor, in *Proc. of the IEEE-EMBS Special Topic Conference on Microtechnologies in Medicine and Biology* (Lyon, France), pp. 607–610.

[25] Oliver, N. and Flores-Mangas, F. (2006). Healthgear: A real-time wearable

system for monitoring and analyzing physiological signals, in *Proc. of the International Workshop on Wearable and Implantable Body Sensor Networks (BSN)* (Cambridge, MA, USA).

[26] Paavola, M. (2007). Wireless technologies in process automation, review and an application example, .

[27] Paavola, M. and Leiviska, K. (2010). Wireless sensor networks in industrial automation, *INTECH, Factory Automation* .

[28] Patel, S., Lorincz, K., Hughes, R., Huggins, N., Growdon, J., Standaert, D., Akay, M., Dy, J., Welsh, M. and Bonato, P. (2010). Monitoring motor fluctuations in patients with parkinson's disease using wearable sensors, *IEEE Transactions on Information Technology in Biomedicine* **13**, 6, pp. 864–873.

[29] Petre, F. and Engels, M. (2009). Wireless sensor networks for industrial appli-
cations, FTMC, http://www.wireless-community.be/Images/pdf/Wireless-Community-FMTC-Marc-Engels-19Feb2009.pdf.

[30] Piran, M. J., Rama Murthy, G. and Praveen Babu, G. (2011). Vehicular ad hoc and sensor networks; principles and challenges, *International Journal of Ad hoc, Sensor and Ubiquitous Computing* **2**, 2.

[31] Ruiz-Garcia, L., Barreiro, P., Rodriguez-Bermejo, J. and Robla, J. I. (2007). Review. monitoring the intermodal, refrigerated transport, of fruit using sensor networks, *Spanish Journal of Agricultural Research* **5**, 2, pp. 142–156.

[32] Sadeghi, M., Khosravi, F., Atefi, K. and Barati, M. (2012). Security analysis of routing protocols in wireless sensor networks, *International Journal of Computer Science Issues* **9**, 1:3, pp. 1694–1714.

[33] Ulrich, T. (2008). Wireless network monitors h2o., wines and vines, http://www.winesandvines.com/template.cfmsection=featurescontent=56594.

[34] Vitturi, S., Carreras, I., Miorandi, D., Schenato, L. and Sona, A. (2007). Experimental evaluation of an industrial application layer protocol over wireless systems, *IEEE Transactions on Industrial Informatics* **3**, 4, pp. 275–288.

[35] Wang, G., Cao, G., La Porta, T. and Zhang, W. (2005). Sensor relocation in mobile sensor networks, in *Proc. of the 24th Annual Joint Conference of the IEE Computer and Communications Societies (INFOCOM)* (Miami, FL, USA), pp. 2302–2312.

[36] Wang, N., Zhang, N. and Wang, M. (2006). Wireless sensors in agriculture and food industry, recent development and future perspective, *Computers and Electronics in Agriculture* **50**, pp. 1–14.

[37] Zhao, G. (2011). Wireless sensor networks for industrial process monitoring and control: A survey, *Network Protocols and Algorithms* **3**, 1, pp. 46–63.

[38] Zheng, Y. (2011). Development of a ubiquitous industrial data acquisition system for rotogravure printe press, *Journal of Networks* **6**, 11, pp. 1543–1548.